ACHIEVING XXCELLENCE IN SCIENCE

Role of Professional Societies in Advancing Women in Science

Proceedings of a Workshop
AXXS 2002

Sally Shaywitz, M.D. and Jong-on Hahm, Ph.D., Editors

Committee on Women in Science and Engineering

Policy and Global Affairs

NATIONAL RESEARCH COUNCIL
OF THE NATIONAL ACADEMIES

THE NATIONAL ACADEMIES PRESS
Washington, D.C.
www.nap.edu

This project was supported by the Office of Research on Women's Health, National Institutes of Health, Grant No. N01-OD-4-2139, Task Order #103, The Burroughs Wellcome Fund, and the National Academy of Sciences. Any opinions, findings, conclusions, or recommendations expressed in this publication are those of the author(s) and do not necessarily reflect the views of the organizations or agencies that provided support for the project.

International Standard Book Number 0-309-09174-8 (Book)
International Standard Book Number 0-309-53104-7 (PDF)

Additional copies of this report are available from the National Academies Press, 500 Fifth Street, N.W., Lockbox 285, Washington, DC 20055; (800) 624-6242 or (202) 334-3313 (in the Washington metropolitan area); Internet, http://www.nap.edu

Printed in the United States of America

THE NATIONAL ACADEMIES
Advisers to the Nation on Science, Engineering, and Medicine

The **National Academy of Sciences** is a private, nonprofit, self-perpetuating society of distinguished scholars engaged in scientific and engineering research, dedicated to the furtherance of science and technology and to their use for the general welfare. Upon the authority of the charter granted to it by the Congress in 1863, the Academy has a mandate that requires it to advise the federal government on scientific and technical matters. Dr. Bruce M. Alberts is president of the National Academy of Sciences.

The **National Academy of Engineering** was established in 1964, under the charter of the National Academy of Sciences, as a parallel organization of outstanding engineers. It is autonomous in its administration and in the selection of its members, sharing with the National Academy of Sciences the responsibility for advising the federal government. The National Academy of Engineering also sponsors engineering programs aimed at meeting national needs, encourages education and research, and recognizes the superior achievements of engineers. Dr. Wm. A. Wulf is president of the National Academy of Engineering.

The **Institute of Medicine** was established in 1970 by the National Academy of Sciences to secure the services of eminent members of appropriate professions in the examination of policy matters pertaining to the health of the public. The Institute acts under the responsibility given to the National Academy of Sciences by its congressional charter to be an adviser to the federal government and, upon its own initiative, to identify issues of medical care, research, and education. Dr. Harvey V. Fineberg is president of the Institute of Medicine.

The **National Research Council** was organized by the National Academy of Sciences in 1916 to associate the broad community of science and technology with the Academy's purposes of furthering knowledge and advising the federal government. Functioning in accordance with general policies determined by the Academy, the Council has become the principal operating agency of both the National Academy of Sciences and the National Academy of Engineering in providing services to the government, the public, and the scientific and engineering communities. The Council is administered jointly by both Academies and the Institute of Medicine. Dr. Bruce M. Alberts and Dr. Wm. A. Wulf are chair and vice chair, respectively, of the National Research Council.

www.national-academies.org

STEERING COMMITTEE FOR AXXS 2002 WORKSHOP

Sally Shaywitz, M.D., *Chair*, Professor of Pediatrics, Yale Center for the Study of Learning and Attention and Yale University School of Medicine, Yale University

Nancy Andrews, M.D., Ph.D., Leland Fikes Professor of Pediatrics, Harvard Medical School

Janet Bickel, M.A., Former Associate Vice President for Medical School Affairs, Association of American Medical Colleges

Michael Lockshin, M.D., Professor of Medicine, Weill College of Medicine of Cornell University

Herbert Pardes, M.D., President and CEO, New York-Presbyterian Hospital

Deborah Powell, M.D., Dean and Assistant Vice President for Clinical Affairs, University of Minnesota Medical School

W. Sue Shafer, Ph.D., Deputy Director, Institute for Quantitative Biomedical Research

Jeanne Sinkford, D.D.S., Ph.D., Associate Executive Director, American Dental Education Association

Project Staff

Jong-on Hahm, Ph.D., Director
Amaliya Jurta, Senior Project Assistant (through July 2002)
Elizabeth Briggs Huthnance, Program Associate

Preface

If the world of biomedical research can be likened to a globe, perhaps clinical research can be envisioned as the side facing away from the sun. Although part of the whole, it is not the shining face of biomedical research. But basic and clinical research share equally the responsibility for achieving the goals of biomedical research—improved health and treatment of disease.

This workshop, held July 8–9, 2002, in Washington, D.C., looked at ways to advance women scientists careers in clinical research. Interest in such careers is falling among medical degree recipients, and particularly among women. This situation is worrisome because, according to the Association of American Medical Colleges, women will soon make up the majority of recipients of medical degrees and life science doctorates (according to NSF data), and declining interest from the growing pool of future investigators may constrict our understanding of human disease.

The Office of Research on Women's Health (ORWH) at the National Institutes of Health asked the Committee on Women in Science and Engineering at the National Research Council (NRC) to hold a workshop to explore ways in which scientific societies could enhance the research careers of women scientists, in support of ORWH's ongoing efforts to promote women's advancement in biomedical careers. Scientific societies play a crucial role in career development, and identifying specific strategies that societies could deploy might be very helpful in encouraging women to enter and continue in clinical research careers. This volume consists of the presentations made at the workshop, including remarks by the leaders of the five breakout sessions. The statements made in the enclosed papers are those of the individual presenters and do not necessarily represent positions of the committee or the National Academies.

ORWH has consistently been a leader on this issue, and the committee would like to acknowledge Dr. Vivian Pinn, director of the Office of Research on Women's Health, for her continued support of efforts to advance women in biomedical research careers, and Ms. Joyce Rudick in ORWH for translating the visions into reality. We also would like to acknowledge Dr. Jong-on Hahm, Amaliya Jurta, and Elizabeth Briggs Huthnance of the Committee on Women in Science and Engineering, for their energetic efforts and commitment in bringing this workshop and the resulting proceedings to fruition.

This volume has been reviewed in draft form by individuals chosen for their diverse perspectives and technical expertise, in accordance with procedures approved by the NRC's Report Review Committee. The purpose of this independent review is to provide candid and critical comments that will assist the institution in making its published report as sound as possible and to ensure that the report meets institutional standards for quality. The review comments and draft manuscript remain confidential to protect the integrity of the process.

We wish to thank the following individuals for their review of this volume: Veronica Catanese, New York University Medical Center; Adrian Dobs, Johns Hopkins University; Elaine Gallin, Doris Duke Charitable Foundation; John Lumpkin, Robert Wood Johnson Foundation; Joan Lunney, United States Department of Agriculture; Christine Seidman, Harvard Medical School; and Harold Slavkin, University of Southern California.

Although the reviewers listed above have provided constructive comments and suggestions, they were not asked to endorse the content of the individual papers. Responsibility for the final content of the papers rests with the individual authors.

Sally Shaywitz, *Chair*
Steering Committee for AXXS 2002 Workshop

Contents

SESSION III: REPORTS OF BREAKOUT SESSIONS

SESSION IV: CLOSING PLENARY

APPENDIXES

SESSION I

Welcoming Remarks and Opening Keynote Address

Welcoming Remarks

Vivian Pinn, M.D.
Director, Office of Research on Women's Health
Associate Director, Research on Women's Health
National Institutes of Health

Let me welcome each one of you to the opening session of our AXXS conference. By now, all of you are familiar with our acronym, the A–double X chromosome–S, for Achieving Excellence in Science. We are pleased about the work that has been accomplished.

One of the mandates of the Office of Research on Women's Health at the National Institutes of Health is to increase opportunities for the recruitment, retention, advancement, and reentry of women into biomedical careers. We were delighted when some of the women scientists at NIH asked us to work with them to increase attention to women's careers through professional societies. It was because of the ideas of people like Sue Shafer (then deputy director of the National Institute of General Medical Sciences, NIGMS), Pam Marino (also at NIGMS), and others that this effort became one of our most successful.

The topic of women in biomedical careers is extremely important for the health of the profession, and for science. For the men in the audience, a lot of things that we want to look at in terms of developing the careers of women actually are applicable to the careers of men. But we do want an opportunity to focus specifically on the careers of women.

At our first meeting, AXXS '99, we mainly focused on basic science professional organizations. So much was accomplished by AXXS '99—not only the Web site and the report, but also recommendations, which are very valuable for the professional societies.[1] Today's keynote speaker, Carola Eisenberg, has

---

[1] The Web site can be found at www4.od.nih.gov/axxs/default.htm. Also see *Achieving XXcellence in Science: Advancing Women's Contributions to Science through Professional Societies,* NIH Publication No. 00-4777 (Washington, D.C.: National Institutes of Health, 2000).

served as co-chair of our task force on women in biomedical careers, has been a charter member of our advisory committee for research on women's health at the National Institutes of Health, and has been working with us since the very beginning. I'm delighted she is here this evening to give us a wonderful kickoff for our AXXS 2002 conference. She will be introduced by Dr. Sally Shaywitz, professor of pediatrics at the Yale University School of Medicine, a member of the Institute of Medicine, and chair of the steering committee for the workshop.

Opening Keynote Address

Carola Eisenberg, M.D.
Lecturer on Social Medicine
Harvard Medical School

SPEAKER INTRODUCTION
SALLY SHAYWITZ, M.D., CHAIR, AXXS STEERING COMMITTEE

One of the pleasures of being here with you today is that I got to know a bit about our keynote speaker. It is a great pleasure to introduce Carola Eisenberg, who is former dean for student affairs and the first woman full dean at Harvard Medical School. Many of you know Dr. Eisenberg for her role in championing human rights as one of the founding members of the Nobel Prize–winning Physicians for Human Rights.

A native of Argentina, Dr. Eisenberg enrolled in the medical faculty of the University of Buenos Aires with an almost exclusively all-male student body. "I didn't have any role models," she recalls. "I had never even met a woman physician. The only thing I did know was that I wanted to be a psychiatrist." After completing training in adult psychiatry, she won a competitive fellowship to study abroad, because there were no programs in child psychiatry in Argentina. In her words, she "came to the States to stay a year, but the year became a lifetime. I met and married a wonderful man with whom I had in short order two children, then and now the light of my life."

Today, Dr. Eisenberg continues to practice psychiatry, teach, and participate in the international humanitarian efforts. She has mentored countless women through the women's groups she started at MIT, Harvard, and the National Academy of Sciences. As she has said, "There is still a machismo attitude, particularly within some fields of medicine. It's changing, but changing slowly." More efforts should be dedicated to opening doors for women and to following in Dr. Eisenberg's footsteps. So it's a real pleasure for me to welcome her today.

DR. EISENBERG
HOW FAR WE HAVE COME, HOW FAR WE STILL HAVE TO GO: HOW WOMEN
SAVED AMERICAN MEDICINE

It's a pleasure to be here. It's a pleasure to see old friends. I'm looking forward to meeting new friends at this workshop.

I am here today in the role of a historian, not a strategist. Strategists will talk tomorrow. If you have read *How the Irish Saved Civilization*, you will understand the subtitle of my speech, "How Women Saved American Medicine." As a historian, I will make three points before putting on the hat of a clinician. First, women established high academic standards for medicine. Second, women maintained those standards when men began to default. Third, women led the fight to enhance the quality of life for physicians. *When*, and not if, we succeed, men will have gender equity for the first time, as well as women.

To my first point, Johns Hopkins University School of Medicine set the standard for American medicine when it admitted its first class in 1893. What you may not know is that the medical school almost did not open. The trustees called a halt when the Hopkins endowment shrunk below the necessary minimum after the stock market crashed. Five Baltimore Quaker women—Carey Thomas, Mary Elizabeth Garrett, Elizabeth King, Julia Rogers, and Mary Gwinn—stepped into the breach. They offered the necessary funds on two conditions. First, women had to be accepted on the same terms as men. Second, a baccalaureate degree and a real knowledge of French and German would be required for admission. Accepting token women was less objectionable than setting high educational standards. Trustees and faculty alike feared pricing themselves out of the applicant market. As Sir William Osler said to Dr. William Henry Welch, "It is likely we're getting as professors those who would never enter as students."

But the women were adamant, and Hopkins took the plunge. Carey, Mary, Elizabeth, Julia, and Mary had elevated the intellectual standards of American medicine for the century and the millennium to come. Harvard followed Hopkins' lead eight years later—that is, in requiring a bachelor's degree; it took another 50 years to admit women. Abraham Flexner put it succinctly in 1910: "John Hopkins graduates in medicine represent the highest quality this country has produced."

As for the second point, nearly a century later women came to the rescue of American medicine again. In the mid-1970s, the number of male applicants to medical schools began to decline. Ten years later, there were fewer qualified male applicants than places in the first-year class. By the late 1980s, there were not enough male applicants, qualified or not, to fill a freshman class. How were academic standards maintained in the face of the massive male default? The answer is straightforward: women constituted a third of the admitted class.

What had enabled women applicants to increase five-fold between the 1960s and the 1980s? Had there been a mutation in the M.D. gene or the X chromosome? I have found no support for that hypothesis. Could there have been a mutation in the admissions process? You bet there was. Was it a random muta-tion? Not on your life. The mutation was directed by the women's movement. Each year during the seventies and the eighties, as more women were accepted, more women applied. Success begot success. Women applicants to medical schools numbered 2,800 in 1971 and rose to more than 17,000 last year.

What is the empirical evidence for the directed change in the admissions process? Affirmative action for men had held steady for two centuries until 1970. For each subsequent year, until the end of the 1970s, the percentage of women accepted exceeded the percentage of women in the pool by 2–3 percent. That edge is statistically significant given the large numbers.

Then a funny thing happened. The admissions edge for women disappeared in 1980, and it was replaced by a two to three preferment for men in the 1980s as the male pool got smaller. Without ever announcing it or acknowledging it, medical schools adopted affirmative action for men. Unless my memory fails me, those men were not dismissed as affirmative action admits. It is odd, isn't it?

Some of you probably object to affirmative action. You will point out that its existence impugns every promotion of a woman by attributing such promotions to affirmative action rather than merit. Yes, some folks do say that. But what about all those men who become professors only because women are not considered? A diverse faculty is a superior faculty, because it is chosen from a larger pool.

My third point is that we are now engaged in a battle for academic norms that acknowledge the importance of family life as a legitimate value. Women physi-cians with children have been leading fuller lives than most of our male counter-parts. Yes, the price has been heavy, but we have not been deterred. We have richer connections with our children and with our parents and, for that matter, with our husbands, when we have them, than men do with their children, their wives, and their parents. We make richer connections with our patients, because we are more in touch with feelings. To the extent we can reduce the endless hours, the competitive atmosphere, and the exclusive focus on personal achieve-ment in academic medicine, we will have created a better world for men as well as for women.

The way scientific research is organized leads to systematic exploitation of trainees. An enormously competitive system forces principal investigators to work themselves and their fellows for long hours. It may surprise you to learn that among postdocs without children, women work more hours than men. Among those with children, women work many fewer. This situation erodes career pro-gression. Citation half-life in the biosciences is short, making it difficult to take time off from work and return to the same career trajectory. Men and women alike need basic reform in the career structure. We need to increase fellowship stipends, to convert lengthy postdoc fellowships into faculty or staff positions, and to provide support for independent research careers at the end of postdoctoral training.

But we must face the fact that women professionals who bear children deal with problems that have no tidy solutions. Even if an enlightened university provides paid parental leave, it cannot authorize leave from the rapid pace of clinical science. Even if a unit chief is sympathetic to a slower pace of work while the children are small, NIH [National Institutes of Health] study section members could not care less. What has she published lately, they will ask, when your grant is up for renewal.

Some women find returning to work after the birth of a baby acutely painful, whether it occurs at 2 weeks, 2 months, or 12 months. With good child care arrangements, the baby does splendidly. It is the mother who suffers separation anxiety, guilt, and loss, sometimes for months. Symptoms recur as women near 40 and begin to wonder whether they want one more child before time runs out.

So those are the three points. Now I put on my clinical hat. During 50 years at three leading universities—Johns Hopkins, MIT, and Harvard—I have been consulted frequently by female colleagues for a gender-specific clinical disorder: reflex alphanumeric narcolepsy, or RAN, as in she RAN all the time trying to keep up. At Hopkins, RAN was known by the eponym "Carola's curse." It is a reflex response precipitated by reading either letters or numbers, thus alphanumeric. The chief complaint is an irresistible closing of the eyes and a rapid descent into slow-wave sleep the moment one sits down in the evening to read a journal.

It strikes on an average day for a doctor who is a mother—that is, on a day that consists of getting up in time to make breakfast for the children, getting them dressed and off to school, rushing to work, seeing patients, leading a seminar, reviewing the NIH pink sheet awarding one a grant score just below the funding level, attending a committee meeting as the token female, squeezing in emergency consultations, rushing home just in time to meet the children returning on the 5:30 school bus, preparing dinner, spending an hour of quality time with the children, getting them to bed, and then trying to read. And suffering, of course, acute narcolepsy. In my view, this was no special day. Neither of the children was sick. There was no call from hubby announcing unexpected dinner guests. Not even a flat tire on the way to work.

My colleagues moan, why can't I stay awake? What is the matter with me? My husband can read until midnight. Well, it doesn't take Francis Crick or Sigmund Freud to understand the pathogenesis of reflex alphanumeric narcolepsy. The condition is neither hereditary nor fatal. Therapy begins with the realization that Wonder Woman is or was a comic book character. The solution lies in renegotiating the division of labor at home.

Life demands choices. You cannot be home full time and at work full time. It's great if your husband takes parental leave from his work. He will be better off for the experience. The children will be grateful to him. I assure you the baby will survive him. But that does not make parting any easier. Many women academics and their children do splendidly with a rapid return to work if partners cooperate, families pitch in, and the couple can afford and find good child care.

As I conclude, I resume my historian's role to review what has been accomplished since our 1999 meeting. You will recall that Nancy Hopkins had just made history. She had assembled the tenured women in the MIT School of Science to protest the inequity in salaries, research space, and departmental governance. When the administration conducted its own study, its findings validated those claims, and it made them public. President Charles Vest called for comparable scrutiny of gender discrimination in the other MIT faculties and found similar data. The institute also took the initiative of convening a meeting of leading universities on gender inequity.

When I became dean of students at MIT in 1972, I was the first woman to sit on MIT's Academic Council, its highest internal governing body. It was a heady place to be, but lonely. Today, 36 years later, six women sit on the council. That's no small progress. Three years ago, not one of us would have imagined that in 2002 four of our leading universities (and this is not the complete list)—Princeton, Michigan, Brown, and Pennsylvania—would have women presidents. At Princeton, President Shirley Tilghman has appointed women to half of Princeton's top academic jobs. And what women! That is worth celebrating.

However, it's far from time to declare victory and send the troops home. If we have won the admissions battle, we are making slow headway, at best, in faculty representation. Yes, there are more tenured women at medical schools than there were five years ago. But the percentages remain substantially below the available pool. Success has been greater in pediatrics, public health, and psychiatry; it has been least in the surgical specialties. To quote from the Association of American Medical Colleges [AAMC] report on increasing women's leadership in academic medicine,

> Few schools, hospitals or professional societies have what might be considered a critical mass of women leaders. The pool of women from which to recruit academic leaders remains small. The potential of most women is being wasted at a time when medicine needs all the leadership talent it can develop to address accelerating institutional and societal needs.

Later in this workshop, Janet Bickel, AAMC vice president and the principal author of that report, will tell us what has and has not been accomplished. I invite you to join me in saluting Janet for her outstanding contribution to the achievement of gender equity.

Have we got it made yet? I conclude with a quote from Estelle Ramey, professor of physiology at Georgetown, who pulled no punches. "Don't tell me we've achieved gender equity when a female Einstein becomes a professor. I'll know we've made it when a female schlemiel is as likely to become a professor as a male schlemiel." I love that quote. My empirical research has identified precious few female professors meeting Ramey's criteria. We have come a long way, baby, but we have a long way to go. For us, for the sake of patients, for the sake of the profession, even for the sake of man, we cannot afford to stand still.

SESSION II

Presentations and Panel Discussion

Opening Remarks

Vivian Pinn, M.D.
Director, Office of Research on Women's Health
Associate Director, Research on Women's Health
National Institutes of Health

I am pleased to welcome you to "Achieving XXcellence 2002" to look at the role of professional societies in advancing women's careers in science and clinical research. I will tell you briefly about how we got into this project.

The Office of Research on Women's Health was established at the National Institutes of Health [NIH] in September 1990, almost 12 years ago, with three major mandates: (1) to determine what we know and don't know about women's health research; (2) to establish a research agenda; to fund, encourage, increase, and stimulate research related to women's health; and to ensure that women are included in clinical studies related to the health of women; and (3) to develop opportunities for the recruitment, retention, reentry, and advancement of girls and women in biomedical careers. Those are the mandates that led us to the topic of this meeting.

Now let me address the third mandate—the recruitment, retention, reentry, and advancement of women in biomedical careers—and how it has gone over the years. As with most of our major efforts at the NIH, we design specific programs with input from the greater public and the scientific community. Some years ago we held a public hearing and a workshop entitled "Women in Biomedical Careers: Dynamics of Change, Strategies for the 21st Century." As I pointed out in the opening session, an NIH task force led this effort and produced a full report of that effort. That task force was co-chaired by Dr. Carola Eisenberg and Dr. Shirley Malcom of the American Association for the Advancement of Science.

At the time of that effort, some of the barriers that affect women and girls entering and succeeding in biomedical careers were identified. Even though these barriers were pointed out a few years ago, they have held over the years as we have continued to review the potential barriers to women entering or advancing in biomedical careers.

In taking these barriers to heart, looking for recommendations from that workshop, and assessing all of our efforts since, including the update of our research agenda, we have begun to put in place career development programs to help our office provide support for girls and women in biomedical research careers and health care professions. In fact, I must compliment Joyce Rudick, who has overseen almost all of our programs related to career development.

But the major one we are addressing today is related to "AXXS, Achieving XXcellence in Science." This idea was first brought to us by Sue Shafer—then deputy director of the National Institute of General Medical Sciences at the NIH—and other intramural women scientists at the NIH. These women asked if we would support them in a joint activity with the American Society for Cell Biology. The initiative would help representatives of professional societies exchange information about strategies they could adopt to support women in their career development. Our first meeting was held in 1999 to bring together representatives of various organizations and professional societies.

We were told not to expect much collaboration or much support from many societies because there would not be great interest. However, to our pleasure and to the amazement of others, our first meeting was extremely successful, with 93 participating organizations. We were exhilarated with the interest and the ambition of the participants, who took that meeting seriously and have continued to work on developing ideas related to AXXS. The wonderful report of the AXXS '99 workshop summarizes its recommendations, and the report can be found on the workshop's Web site [www4.od.nih,gov/axxs].

As we've moved from 1999 to AXXS 2002, there have been more efforts by various working groups to address mentoring and networking, career development, achieving senior and leadership levels, representation of women in scientific societies, and developing and identifying model systems that work. These groups have also been pursuing outreach and collaboration within and between societies and organizations.

This brings us to AXXS 2002, which is where we are today, to look at the role of professional societies in advancing women's careers in science and clinical research. We turned to the Committee on Women in Science and Engineering at the National Academy of Sciences to assist us and take over this effort and move it forward in 2002. We asked that the committee develop a workshop at which clinical societies could come together to discuss ways that societies can enhance the participation of women scientists in the clinical research workforce. And we requested that the workshop focus particularly on initiatives and action items that clinical societies can adopt within their organizations to enhance women's advancement in the clinical research field; on ways that clinical societies can disseminate proven, successful strategies in order to advance women's careers; and on ways that clinical societies can collaborate with each other to promote women's contributions to their fields.

Sally Shaywitz, M.D.
Professor of Pediatrics, Yale University School of Medicine
Chair, AXXS Steering Committee

As chair of the steering committee, I'm delighted to add my welcome to the AXXS 2002 workshop, "Achieving XXcellence in Science: The Role of Professional Societies in Advancing Women's Careers in Science and Clinical Research." This workshop aims to expand the progress achieved at AXXS '99, which gathered representatives of basic science societies to discuss ways to encourage women scientists. We are tremendously excited by the potential of this workshop to bring about change and to enhance the participation of women scientists in clinical research, and to do so through the efforts of our clinical societies.

We strongly believe that professional societies can make a difference. They play a key role in career development. They can appoint women to editorial boards, to important committees—for example, to nominate candidates for awards—and to positions as committee chairs and speakers. In essence, professional societies often provide the currency that counts in advancement in academic medicine.

Our goal today is to determine how the important role of societies can be leveraged to enhance women's advancement in clinical research. We have defined an ambitious agenda—one that will allow us at the close of this workshop to identify specific initiatives and activities that can be adopted within our societies, to develop mechanisms to disseminate successful strategies to advance women's careers, and to determine ways that clinical societies can collaborate to promote women's contributions to their fields.

At the onset, let's also clear up some incorrect assumptions. It was once thought that the underrepresentation of women in leadership roles in clinical research was mainly a pipeline problem—that is, not enough women were entering medicine and so fewer women were available in the pool for selection to academic leadership positions. This is no longer a tenable explanation. As you will hear later, women are entering medical school in numbers almost equal to those of men.

Furthermore, in the past proportionately more women medical school graduates than men chose academic medicine as a career pathway. This appears to be changing now that fewer women seem interested in choosing academic medicine. Instead, proportionately more women appear to be seeking careers in private practice or industry.

Throughout all of this, what has not changed is the underrepresentation of women in the ranks of senior faculty. Women are obtaining faculty appointments and increasing their representation on medical school faculties. But these are junior, not senior, positions. The proportion of women faculty who are full professors has not changed in over 15 years. According to an editor's note in the *Journal of the American Medical Association*, "Even if the rate of women attaining full professor rank continues to grow yearly, at least twenty-five years remain until the proportion of women at full professor rank is half that of men, despite near gender equity when entering medical school."

So as we examine the chain of academic leadership, we note that women are represented in fewer numbers as department chairs or as members of important committees—those that wield power rather than those that take care of the housekeeping items. At the top of the chain, according to the AAMC [Association of American Medical Colleges] database on medical school deans, only 4.1 percent of all deans of U.S. medical schools are women.

But just as the problem is becoming increasingly evident, so are some of the solutions. For example, there is general agreement about the essential role of mentoring in advancing the careers of both men and women. Unanswered questions concern the fewer numbers of senior faculty women available to assume mentorship roles and the possibility that some approaches are more successful than others for women. Women often seek collaborative approaches and that brings with it the potential for exploitation.

These issues are now on the table for clinical societies to address, and, optimally, to help resolve. As editor Cathy D'Angelis asked in an editorial in the *New England Journal of Medicine*, we now ask: What can societies do to promote effective mentorship? How can societies go about promoting seminars and sponsoring discussion groups that help women in negotiating research time, space, and institutional resources? How can societies help clinical researchers become cognizant of the importance of a sharp, circumscribed academic focus?

Our focus today is not so much to convince you of the need, but to develop a workable action plan to bring about change. Our goal is to transform the advancement of women in clinical medicine into a central issue rather than a peripheral or side one. The data are convincing, but we want more than to just document the problem. We want—in fact, we must create—change. Such data are not endpoints. Rather, they serve to persuade us of how much must be done.

This is not a symposium but a workshop, aptly named because we're asking each of you here today to work with us to problem-solve and to take potential solutions back to your societies and act on them. We hope to be following the results of these actions through tracking programs over the next several months and, we hope, years. This work is important for each of you and for your societies, and for the community of clinical research scientists and for our patients and for society at large. The increased representation of women scientists as leaders in clinical research will be good for individual women, for their institutions, and for society. What will be good for women will be good for all.

Indeed, the problems that clinical research addresses are too important, too complex, and too elusive to waste any talent that might provide new insights, new ideas, or new approaches. We must ensure that the makeup of the leadership of clinical research efforts resembles that of the society we are serving, nothing less. I urge you to be bold in your recommendations. There is an urgent need for not one token woman but at least several women to be appointed for each promotion and search committee, including those for deans and presidents. It can be done.

In 2001 Shirley Tilghman became president of Princeton University. To quote a recent *New York Times* article, "When she became president, Dr. Tilghman said she knew how to close the lingering gap. At Princeton, women made up about 27 percent of the faculty but only 14 percent of full professors. The key, she said, was to appoint more women as administrators. And she did—as provost, as dean of the Woodrow Wilson School, as dean of the School of Engineering and Applied Science, and as dean of the undergraduate college—a total of five women among the nine top academic officers. Such bold actions do not sit well with everyone. Recalling that Princeton was once an all-male school, U. M. F. Lewis, a Princeton alumnus, class of '41, wrote in the *Princeton Alumni Weekly*, "To save time, I recommend that the trustees promptly convert Princeton into a single-sex, female university, and be done with it." In a subsequent issue, a 1993 graduate, Betsy Helman, wrote: "Based on your letter, Mr. Lewis, it is clear that you are no tiger. You are a dinosaur."

So Princeton demonstrated that change is possible and that it can come about quickly. According to Nancy Hopkins, professor of biology at the Massachusetts Institute of Technology and a leader in bringing issues of gender inequality in academia to the fore, "Having women in power sends a message to young women that, yes, of course, you can become president of a university, win the Nobel Prize, or do anything. Up to now we've been telling them that, but no one was *showing* them." Our mission, then, is to figure out how to show women at all levels of clinical research that they too can be president, can win a Nobel Prize and certainly can be professor.

As we begin our formal program for the workshop, I wish to point out that this is an auspicious occasion on several fronts. As far as I know, it marks the first time the Institute of Medicine [IOM] has formally addressed the issue of advancing women scientists' careers in clinical research. We are delighted to acknowledge the participation of the IOM Clinical Research Roundtable. And we are especially pleased that the new president of the IOM has signified his interest and his support by joining us in welcoming you to what we believe will be a landmark event in the progress of women scientists engaged in clinical research.

Harvey Fineberg, M.D., Ph.D.
President, Institute of Medicine

SPEAKER INTRODUCTION
SALLY SHAYWITZ, M.D., CHAIR, AXXS STEERING COMMITTEE

It is a great pleasure to introduce Harvey Fineberg, who on July 1, 2002, became the seventh president of the Institute of Medicine [IOM]. Dr. Fineberg earned his bachelor's degree from Harvard University, his medical degree from Harvard Medical School, and his master's and doctoral degrees in public health from Harvard University's Kennedy School of Government and Graduate School of Arts and Sciences, respectively. He most recently served as provost of Harvard University and, before his appointment as provost, as dean of the Harvard School of Public Health.

Dr. Fineberg was elected to the Institute of Medicine in 1982, but his work for the institution dates back 25 years. He has chaired various important committees of the National Academies, including those that produced the reports *Understanding Risk: Informing Decisions in a Democratic Society* and *Society's Choices: Social and Ethical Decision-Making in Biomedicine*. His wide-ranging research interests encompass HIV/AIDS and other infectious diseases, the fields of risk assessment and decision making, the evaluation of diagnostic tests and vaccines, the ethical and social implications of new medical technologies, and medical education.

In announcing his appointment, National Academy of Sciences President Bruce Alberts said, "Dr. Fineberg's background and skills are ideal for this position. Public health has become recognized as an area of increased national importance, which will make IOM's mission to advise the nation's health policy even more critical." Kenneth Shine, Dr. Fineberg's predecessor as president, noted, "Harvey Fineberg combines a rich academic leadership experience with a continuing commitment to and involvement in the health of the public." He is an outstanding choice, and I am so delighted to be able to welcome him and introduce him to you.

DR. FINEBERG:

It is a pleasure for me to be here with you at this workshop for many reasons. First, it's just a delight for me to be established now here at the Institute of Medicine, within the National Academies. This is actually the first workshop or working program that I've been privileged to welcome since my appointment just a week ago. I could not imagine a more fitting way to begin as the president of the Institute of Medicine.

When I saw the title of this enterprise, I actually thought the pronunciation was double-excellence. The reason that seemed especially meaningful to me is that it's so obvious that society cannot afford to squander half of the scientific and clinical brainpower available to us. Aside from how important it is to individuals, the purpose of this activity to me from a social point of view is very simple: we need to take fullest advantage of every individual's talent and ability to contribute. But we're failing to do that. We're failing to do it for women. We're failing to do it for disadvantaged minorities. We're failing to do it for reasons that have nothing to do with that individual's ability to contribute.

I hope that in the course of your deliberations each of you representing a professional and clinical society can carry back two or three really good ideas that you had not previously thought of. If each society can introduce just a few initiatives that can make a difference, I believe the effect will be cumulative and significant.

This is a field for long-distance runners; it is not a place for sprinters who run out of breath. This is a field that requires perseverance. At the same time, I don't think we have to content ourselves only with distant and remote solutions. I believe there can be positive tipping points. I think Princeton is a good example. Believe me, in academia if women can take over at Princeton, they can take over anywhere.

I hope we will find a way to put those initiatives, those new activities, and those commitments into place, so they can bring more of our organizations, institutions, and societies to that positive tipping point where the place of women is no longer a matter for future solution but a matter of current reality for scientists, clinicians, and leadership at every level. I commend each of you for being here, and I wish you every success today and in the months and years to follow.

Keynote Address
Women in Science and Medicine

Karen Antman, M.D.
Columbia Presbyterian Medical Center

SPEAKER INTRODUCTION
SALLY SHAYWITZ, M.D., CHAIR, AXXS STEERING COMMITTEE

I'm also absolutely delighted to introduce our keynote speaker, Dr. Karen Antman. Dr. Antman, director of the comprehensive cancer center at Columbia Presbyterian Medical Center and the Wu Professor of Medicine and chief of medical oncology, has developed a number of now-standard regimens for the treatment of certain forms of cancer, has developed high-dosage chemotherapy regimens for high-risk breast cancer, and is testing various strategies of bone marrow or stem cell transplantation to replace immune stem cells lost during high-dosage chemotherapy. Dr. Antman is one of only four women currently serving as director of one of the National Cancer Institute's 59 designated cancer research centers. So it is with great pleasure and enthusiasm that I introduce you to Dr. Antman.

DR. ANTMAN:

It's a privilege to be asked to give this talk.

The data I will show you during this presentation are not from a Medline search of what's been published; it is basically the same kind of literature review one would get as an academic from ripping pages out of journals over a period of about 20 years.

I received the following as an e-mail; it was kind of a joke. The hypothesis is that success in academia requires specific phenotypes. The abstract read:

21

We used human clones to identify the molecular events that occurred during the transition from a graduate student to a professor. A pool of graduate students was selected on minimal money media and they were dubbed post-docs. These were further screened for the ability to work long hours with vending machine snacks as their sole carbon source. Those selected by their ability to turn esoteric results into a fifty-minute seminar were labeled assistant professors. Clones which overproduced stress proteins, heat shock protein 70, were passed over "friends and family members" columns, and such selected full professors shared striking phenotypes: the inability to judge the time required to complete bench-work and the belief that all of their ideas constituted good thesis projects. Over-expression of these selected gene products may speed evolution of graduate students to full professor.

The point of this is that making a contribution in science and medicine is not easy for either men or women. But there are differences. When women are depressed, they either eat or go shopping. Men invade another country. At many universities, as you'll see in the next talk, the numbers of instructors and assistant professors are roughly equivalent to the pool since the 1970s. The percentage diverges with each promotion.

Failure of the Trickle-up Theory

Why hasn't the trickle-up theory worked? I've heard lots of people commenting, both women and men. Is it commitment, frustration on the part of women, discrimination, or small differences in early resources? I think the latter reason is one that very subtly contributes to the differences and probably is part of the problem, although I believe the others are as well.

A Columbia University commission on the status of women reported out in October 2001. It found that 40 percent of the Ph.D.'s in arts and sciences were given to women. However, when it came time to do academic appointments for tenure-track jobs, 23 percent were in the pool. They said there was a puzzling absence of qualified women. I thought the choice of words there was interesting. (It's also interesting that, at Columbia at least, the hiring was actually done at 34 percent. So they actually hired a higher percentage of women than was in the applicant pool.)

The commission considered the reasons for fewer women in the applicant pool, and one was New York City. Did women opt out of coming to New York City? If women are perfectly willing to get their Ph.D.'s in New York City, it doesn't make sense to me that they would not be willing to take an academic appointment in New York City, but that was one of the reasons given: "Advising networks might steer women away from elite institutions or suggest a career at a research institute is incompatible with commitments to raising small children." Which I thought was an interesting concern. The report also said, "Women may underestimate their qualifications."

Let's look at the latter two concerns in a little bit more detail. Family responsibility is an issue. The time commitments are impressive. We have two children (actually, they're both now in medical school), and the amount of time required to care for and raise them was really an issue. I think women have to trade money for time, but when I've said that to graduate students and postdocs, they don't have money at that point in their careers. So just about the time they're having children and need adequate child care and a car and an apartment, they don't have the money to get the basic necessities. This is particularly a problem in New York City.

As for whether women may underestimate their qualifications, what might be the evidence for that? An interesting Project Access was partially reviewed in *Science* in 1996.[1] The authors interviewed a variety of men and women in medicine and science in academic medical centers, and found that 70 percent of the men thought they had above-average ability compared with 50 percent of the women. So I believe it's fairly clear that women are less likely to be confident about their ability in science.

Twenty-five percent of the women, versus 5 percent of the men, said that, in retrospect, they should have dealt more actively with career obstacles. But it's not always easy to do that when you're a young assistant professor and don't really know how to deal with these obstacles. So I think that senior women and men really have to provide ways for women to reduce the obstacles to the development of their careers.

The authors also found that women were slightly more collaborative prior to their postdoc, but collaborated noticeably less thereafter, presumably because men were not treating them as equal partners. I think this is worrisome—that early on they collaborate and then they find that they have more difficulty collaborating on an equal basis as their careers advance.

One of the issues that did come up in Project Access about promotions was that the men interviewed published slightly more papers per year (2.8) than the women (2.3). However, when the analysis was expanded, using a citation index, it was learned that the women had strikingly higher citation indexes than the men per paper. I believe this factor needs to be taken into consideration. The authors concluded that the women were more cautious and careful about adopting an extra-high measure of conformity to research formalities.

Finally, it was discovered that women postdocs—and this is particularly worrisome to me—with female advisers left science more often than those with male advisers. The reason they gave was that the female adviser had given up any personal life and that they didn't want to do the same. We can't solve this problem by getting more female advisers who have given up everything. We have to have reasonably well-rounded female advisers who can serve as role models,

---

[1] "Women and Minorities," *Science* 271 (1996): 295.

because junior women coming up through the ranks don't want unilateral kinds of role models. In fact, I remember, as an assistant professor, complaining to one of the administrators that there were no normal women role models. She said, "Look around—there aren't a lot of normal men advisers either." This is a problem for young men too—not having enough time to spend with their families.

Different Cultures, Different Rules

I believe that the cultures and the rules are different for men and women. Many of you may have read Carol Gilligan's work from about 10 years ago, *In a Different Voice.*[2] Basically, she was evaluating Radcliffe and Harvard undergraduates as part of her Ph.D. research. She found that the men were always ranked higher on ethical scores than the women. She finally concluded that the women weren't ranked higher on ethical scores because they had different ethics—that is, they had been brought up somewhat differently and had a different set of rules. The men viewed law as important, and women did too, but, for them, the law was that relationships come before the law. Therefore, the women kept getting dinged on that particular basis and were not considered as well developed ethically. By publishing *In a Different Voice*, I believe Gilligan established that there are just differences in the rules that women and men follow.

Promotion at academic medical centers is generally based on independence. Women tend to be collaborative. Many of the junior women whom I've mentored did not realize that if a full professor was listed as coauthor on their papers they would not be considered independent. Full professors know this and should take their names off papers when their colleague becomes a fairly senior assistant professor. Some do and some don't. But young women don't realize that that is the major value system in academic medicine. I'm not sure why collaboration shouldn't be.

Finally, critical mass is essential. I think the "old girls' network" has been very helpful in getting women promoted and in making differences on committees. About 10 years ago I served on the board of the American Society of Clinical Oncology. There were three women and twenty men. We were very aware that often when a woman makes a good suggestion it is dropped and then a little while later is attributed to the men around her. So we three women met ahead of time, and for the whole year whenever one of the women in the group made a suggestion we thought was particularly smart, we would reinforce it. I don't think the guys ever figured out what we were doing, but it was very effective for the year. I think that one or two women can't do it. You really need a critical mass of women.

---

[2]Carol Gilligan, *In a Different Voice: Psychological Theory and Women's Development* (Cambridge, Mass.: Harvard University Press, 1993).

There are no real barriers, I believe, for a brilliant woman faculty member. Getting her a tenured position is not an issue. However, at least in my experience, plenty of B+ men also have gotten tenure, but trying to get a B+ woman tenure *is* often an issue. You're caught in a position of supporting someone who isn't perfect. It's often difficult to get that person promoted.

In 1997 a study came out in *Nature* (and I'd love to see something like this done at the National Institutes of Health—just a quick review to see if in fact this is correct).[3] In 1995 the Swedish courts made public a group of grant applications to the Swedish Medical Research Council. The applications were made by 62 men and 52 women, so it was almost even. But 16 men and 4 women were awarded grants. After looking at these results, the investigators did a multiple regression analysis of the number of first author and total papers; journal impact factor—so they factored in where the applicants were publishing; citation index: education; rank of the applicant's medical school; mentor, field; and postdoc, whether a postdoc had been done abroad. This study found that the women needed about two and a half times the scores of the men to receive the same evaluation. That's worrisome. Many women say they have to be twice as good as men to be considered half as competent. These data support that evaluation.

Women in the Classroom

We all do teaching and we need to be aware of what's happening in the classrooms when we teach. A 1985 paper by Krupnick[4] looked at teaching styles in a Boston classroom and then the classroom discussion. Krupnick found that men dominated mixed-group discussion groups, that women were interrupted far more often than men, and usually by women, and that long periods of all-male talk were followed by short bursts of all-female talk. I've gone to lots of conferences, and in the question period this scenario almost always plays out exactly as described. Therefore, if you're teaching a class and you're a female and you're aware of these data, you probably should be calling on women occasionally in the early male-dominated discussion, just to start mixing it up a little bit. Then when women start interrupting women, you probably need to interrupt that pattern and, say, let the last speaker finish. It's almost as though there's a rush to get it all in once it's the women's turn. We need to allow women to express themselves fully.

And there are still leadership barriers. There's discrimination, but it's both for and against. Some people discriminate against women, but some others very clearly discriminate for women. One of the major problems is that leaders are comfortable with people like them. Since many leaders are men, they're more

---

[3]A. Wold and C. Wennerås, *Nature* 387 (1997): 341–43.
[4]C. G. Krupnick, "Women and Men in the Classroom: Inequity and Its Reminders," *Journal of the Harvard-Danforth Center* (1985): 18–25.

comfortable with men. That's kind of a subtle effect. Assertiveness is often an asset in a man, but it may be problematic in a woman. I believe that when men work for a woman, they very quickly will try to go over her head to the next man. Whether they succeed depends on the sophistication of the woman's boss.

Professional Societies

How can professional societies help? An example is the one I will be leading as its president next year. The American Association for Cancer Research [AACR] has a membership of about 12,000 men and about 5,000 women; 68 percent are Ph.D.'s and 24 percent are M.D.'s. Like in academic medical centers, women make up a higher share (41 percent) of associate members (those who do not yet have academic appointments), almost double that of active members (see Figure 1). So it certainly looks like the pipeline is full.

The AACR has had a Women in Cancer Research Council for decades that is made up of 1,491 women, almost half of the women in the AACR, and 113 men (it's kind of like the Marines, a few good men). Many more people come to the lectures and workshops than are actually in the group. There are breakfasts specifically for the trainees, mentoring programs for both trainees and mid-level

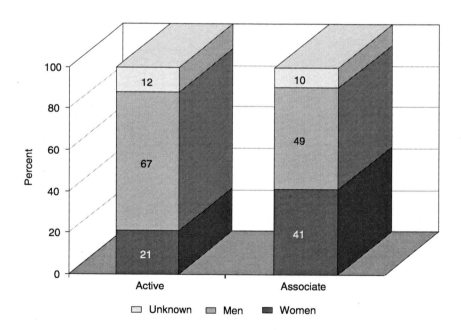

FIGURE 1 Membership of American Association of Cancer Research, by gender. SOURCE: American Association of Cancer Research.

faculty, and scholar awards for women. It is very important that professional societies undertake these kinds of programs to ensure that women have a place to go to get mentoring at both early and mid-career levels.

Questions and Answers

Participant: I was curious about evaluating the programs of the professional societies. Do you have measurable standards to determine how things have improved relative to having breakfasts and mentoring programs, or do you do a survey now and three years from now to really understand the impact of which programs are most effective?

Dr. Antman: Surveys are done. Women are certainly increasing both in their percentages within the society and in their leadership within the society, which is fairly equitable. I'm referring to the number of women on the board, the number of women presidents, and the number of women editors of the five journals that AACR sponsors. There's been a major shift over perhaps a decade and a half. Whether that has to do with this particular program, I don't know how one would tell. But this certainly is a society in which women are well represented at all levels.

From AXXS '99 to AXXS 2002

Page S. Morahan, Ph.D.
National Center of Leadership in Academic Medicine

SPEAKER INTRODUCTION
SALLY SHAYWITZ, M.D., CHAIR, AXXS STEERING COMMITTEE

Dr. Page Morahan will bring us up to date from AXXS '99 to 2002. Dr. Morahan is co-director of the Hedwig van Ameringen Executive Leadership in Academic Medicine Program for Women at Drexel University School of Medicine. And she served as founding director of the National Center of Leadership in Academic Medicine at the Medical College of Pennsylvania. The goal of the center is to develop and implement mentoring programs within academic medicine in order to foster gender equity in medicine and promote the advancement of both women and men junior faculty into senior faculty leadership positions. The center at the Medical College of Pennsylvania is one of four centers sponsored and supported by the Office of Research on Women's Health at the National Institutes of Health [NIH].

DR. MORAHAN:

I've had the pleasure of being involved with this effort since the beginning of the AXXS '99 meeting and look forward to bringing you up to date.

The Office of Research on Women's Health and 20 other NIH entities sponsored the first meeting in 1999. Participating in that workshop were a hundred people from a hundred different societies, but few clinical societies. One of the reasons for this meeting today is to bring on board the clinical research societies. In AXXS '99 we proposed 14 different initiatives within four themes: (1) leadership, visibility, and recognition of women; (2) mentoring and networking; (3) effective practices; and (4) oversight, tracking, and accountability.

Leadership, Visibility, and Recognition of Women

Four of the initiatives under this theme follow the work of Robin Ely and Deborah Myerson who laid out the best strategies to increase numbers of women.[5] The first, the "fix the woman" strategy, gives women skills they may not have. Leadership programs and mentoring programs fall under this approach. The second strategy is to value the feminine—in other words, value the different skills and emphases that women bring to research, and validate them by increased recognition and visibility. The third approach is to create equal opportunity. That would include issuing report cards, for example, on how many women are selected for committee memberships and ensuring equal access to these types of positions.

These three strategies are very important, but they are not sufficient. They increase the numbers of women, but they do not change the fundamental playing field. This is where the fourth strategy comes in—to assess and revise the work culture. In fact, this is part of the effort of AXXS: to create an umbrella organization to examine strategically ways to change the culture of scientific societies so that women's contributions will be more valued. This effort has been spearheaded by Sue Shafer and a coordinating group that was part of the original AXXS planning team. This has been an excellent approach to keeping an effort going and producing some important initiatives that this workshop can now build on.

Mentoring and Networking

Our first major effort under the second theme, mentoring and networking, was to create the AXXS Web site, which now averages about 400 hits a day. It is being further developed as a resource for women in science who are searching for publications and Web links. We'd like for this Web site to serve as a clearinghouse for information on women, science, and strategies for success.

Another major initiative has been to establish an "effective practices" clearinghouse. The first effective practices that we gathered—organizational practices to advance the careers of women in science—are available on the AXXS Web site, and we hope to add more after this meeting. So we challenge all of you and your societies, such as the AACR, to send in practices, particularly those that are different from ones already up on the Web site. We hope that you will steal the ideas of everyone else and use them in your societies.

---

[5]Robin Ely and Deborah Myerson, *Research in Organizational Behavior* (New York: JAI Press, 2001).

Oversight, Tracking, and Accountability

The fourth theme—oversight, tracking, and accountability—requires, we believe, an oversight organization. The Association for Women in Science, or AWIS, is a very useful and powerful organization, but it does not address primarily leadership issues. In academic medicine and dentistry, there is the Society of Executive Leadership in Academic Medicine, but it doesn't include all of the other sciences, math, engineering, and technology. We believe there is a need for an umbrella organization.

It also is very important to establish a report card on the status of women in science and engineering. This has been an exceedingly useful approach in business. Catalyst (www.catalystwomen.org) is a research organization that does an annual report card on the number of women in top leadership positions in Fortune 1000 companies in the United States and Canada and on the boards of those companies. It is highly public, highly publicized. Catalyst has really moved the competitive spirit of corporations and made the business case for the importance of having women on corporate boards.

The Association of American Medical Colleges has used the same approach, publishing an annual report card of medical schools. These days, deans and medical schools don't like to be known as being down on the bottom. They'd much rather be known as one of the top 10 schools for the number of women chairs and division chiefs.

And there have been the ripple effects. These are positive, unintended consequences from starting an effort like AXXS. We need to remember the importance of these effects. They may not be in "the strategic plan," but they are important outgrowths. In the last round of the National Science Foundation's ADVANCE awards were four that meshed very closely with AXXS priorities. Two were the collaborative efforts taken on by the American Chemical Society and the Gordon Research Conferences. And the Gerontological Society of America has begun tracking membership by gender, which it had not done before. Finally, the group Women in Cancer Research has initiated a mid-career mentoring workshop. All of these may seem like small projects, but each of them sends out ripples that can make a difference over the years.

These ripple effects are discussed in Deborah Myerson's book, *Tempered Radicals.*[6] This book shows the importance of small effects by people who choose to work within the system. I call us people who rock the boat but not enough to be dumped out. There is a level of progression from the small activities that we can do, such as supporting flexible work arrangements for the people in our laboratories or our clinical units, on to larger, more organized efforts. That's what we hope to develop today, so I look forward to seeing what will come out of this workshop.

---

[6]Deborah E. Myerson, *Tempered Radicals: How People Use Difference to Inspire Change at Work* (Boston: Harvard Business School Press, 2001).

Questions and Answers

Participant: I'm Joanne Kaufman, executive vice president of the American Society of Human Genetics—the first executive vice president of the American Society of Human Genetics. On the ripple effects, I would just like to add that we should seriously look into additional ways to work within the national and other umbrella organizations rather than create a separate organization. I would suggest collaboration with the Federation of American Societies of Experimental Biology on the basic science side. Another hat I wear is genetics representative to the American Board of Medical Specialties, which is certainly a bastion that needs to be changed. It has a General Assembly of about 120 voting members, and the last time we met 13 women were voting in the assembly.

Which brings me back to one other point, and that is home institution–based rather than just society-based initiatives. Of those 13 women, three of us were from the University of Maryland. Two of us happened to be in surgical specialties. The nurturing from our home institution allowed us to step forward nationally in ways that were very useful. So I agree with you wholeheartedly on working within the system.

Dr. Morahan: In talking about the universities and societies, it really is necessary to work both ways. Sometimes when women become more visible in societies, they are more likely to be tapped for something in their university. Then the reverse can happen, as you described at the University of Maryland. So both ends need to be addressed.

Participant: I'm Roberta E. Sonnino, associate dean for women in medicine and special programs at the University of Kansas. By the way, I'm also representing Dr. Deborah Powell, a member of the steering committee, to some degree. And I'm also here on behalf of the Association of Women Surgeons. To follow up what was just said, one of my concerns that I hope will be answered here is that I know three or four of the groups represented at this meeting are already kinds of sub-branches. The Association of Women Surgeons that I'm representing is a good example. We're already a group of women who are trying to do something. Obviously, I represent one of the specialties that needs help the most. Kimberly Ephgrave is here from orthopedics. Are there other surgeons in the group, or are we it? No, we're it. That was my fear.

Listening to you speak, Page, I realize that we really need to get to the *mainstream* organizations. I'm not saying that we're not mainstream, but when I report back to my executive council I'm preaching to the choir. So I'd like to encourage everybody to help me. How do I get to the American Surgical Association, the American College of Surgeons—all the places where we don't even manage to get our toe in the door, let alone convince them to put up a mentoring workshop for women? Kim and I are in the male-dominated specialties. We're very much looking for help from the group.

Participant: I'm Mahin Khatami, president of Graduate Women in Science in Bethesda, Omicron chapter. I'm very glad you mentioned the importance of leadership for women. I think the impediments to professional women who want to achieve senior intellectual positions within a society, within government, or within universities can be very serious and are an important factor that needs to be considered within the system.

Nobody wants to rock the boat and end up in the sea, but when a woman is competing for seniority in an intellectually sensitive situation, the backlash against her and the retaliation against her can be very serious. The solution I have for this type of situation could be considered an intellectual protection committee.

Dr. Shaywitz: Important issues have been raised on a number of points. One is that all women are not the same and that we do need to look at the different needs of various groups of women.

Participant: I'm Deborah Carper, chair of the NIH Women Scientist Advisory Council. Dr. Morahan, I particularly wanted to emphasize one of the action plans included in your presentation: to develop the *database of women scientists.* At the NIH we just finished a report on leadership positions. Not surprising to us, we found that 40–50 percent of entry-level positions, the training grants, were occupied by women, and that as few as 5 percent of tenured positions were held by women. In particular we need to be able to recruit and to promote women scientists, and bring them in on search committees. To do that, we could compile a national registry or a database of women scientists, so that we could quickly go to this list. As it stands now, when we do a search or we're asked to put women on a scientific council at the NIH, the women in each institute have to come up quickly, sometimes by the end of that day, with a list of 12–15 women who could serve on these search committees and boards of scientific counselors.

It's imperative for us to consider in our action plan compiling a list of women scientists nationally and internationally, so that not only the women but also people working in the administrative areas can have access to a list of women scientists who would be willing to serve on these important positions that lead to our attaining senior leadership roles.

Dr. Morahan: I couldn't agree with you more that we need higher visibility and a better network of women. A database is one way to help do that.

A Pathways Model for Career Progression in Science

Pam Marino, Ph.D.
National Institute of General Medical Sciences
National Institutes of Health

SPEAKER INTRODUCTION
SALLY SHAYWITZ, M.D., CHAIR, AXXS STEERING COMMITTEE

It's my pleasure to introduce Pam Marino, program director for the division of pharmacology, physiology, and biologic chemistry at the National Institute of General Medical Sciences, or NIGMS, at the National Institutes of Health. In this role, Dr. Marino is co-director of the NIGMS intramural Pharmacology Research Associate, or PRAT, program. She directs NIGMS extramural programs in glyco-biology and molecular immunology and serves as the NIGMS liaison to the Office of Research on Women's Health at the NIH. She is a member of the American Association of Cancer Research and serves on the AACR education committee. Dr. Marino is also a member of Women in Cancer Research, in which she has served as co-chair and chair of the mentoring committee.

DR. MARINO:

I want to thank the National Academies' Committee on Women in Science and Engineering and the Office of Research on Women's Health at the NIH for inviting me to speak. Because I was trained as a pulmonary biochemist, I think in terms of biochemical pathways. Quite honestly, I don't like pipelines. They are linear things in which something is put in on one end and something else is expected to come out of the other end. As for "leaking," I can't think of people in terms of applying tape. I think in terms of dynamic systems.

33

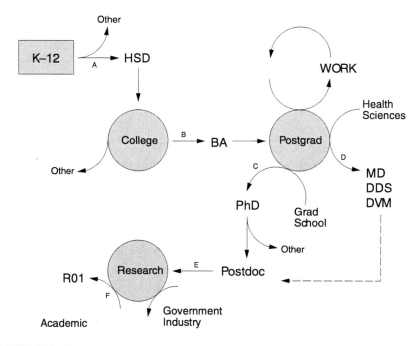

FIGURE 2   Pathways model.
SOURCE: National Institutes of Health, Office of Research on Women's Health, *AXXS '99, Achieving XXcellence in Science, Advancing Women's Contribution to Science through Professional Societies*, Marino, "A Pathways Model for Career Progression in Science," pp. 15-19.

I set up a pathways model for taking people from cradle to grave, using rate constants to describe how we get people through those steps (see Figure 2). But how do we keep people on the pathway and keep them moving forward through that progression? In this presentation I'm going to take each of these steps and talk a little bit about what happens at them in terms of national numbers and women.

Looking at the Numbers[7]

We send kids into schools and we spend lots of money telling them that they should go into science. When we look at the numbers we see that 52 percent of

---

[7]The following material is drawn from: National Institutes of Health, Office of Research on Women's Health, Bethesda, MD., 1999; *AXXS '99, Achieving XXcellence in Science, Advancing Women's Contribution to Science through Professional Societies*, Marino, "A Pathways Model for Career Progression in Science," pp. 15-19.

the population is female. In the United States, we're producing about 2.5 million high school graduates every year, of which 51–52 percent are female. We then send a percentage of them on to college, and we hope they go on to get a bachelor's degree.

A very small percentage of those who do go on to college actually go into biology and chemistry—the major sciences that will supply medical school applicants. In 1981, 44.5 percent of the 44,000 B.A.'s awarded in biology went to women, and 30.1 percent of the some 11,500 B.A.'s awarded in chemistry went to women. By 1996, these numbers had gone up to 53 percent of 62,000 in biology for women and 41.5 percent of 11,000 in chemistry for women. The numbers in biology fluctuate from year to year, but they're between 40,000 and 60,000 over about 20 years. For chemistry, the number of B.A.'s graduating every year is pretty consistent. In 2000, 59.5 percent of the B.A.'s awarded in biology went to women. Sixty percent is a pretty hefty number when you realize that is the pool from which medical school applicants will be drawn. Once the 60 percent is put into the postgraduate pool, they go to work or they go to graduate school or they go to professional schools. Most students seem to sit in this work pool for about two years before they go on to graduate school. Sixteen thousand of them go on to medical school. The percentages of women going in either of these directions are increasing over time.

In 1981, 15 percent of the Ph.D.'s earned in chemistry were earned by women. By 1996, 30 percent of about the same number of degrees went to women. In 2000, 31.4 percent of the chemistry Ph.D.'s went to women. In biology, 29.1 percent of the 3,400 Ph.D.'s produced in 1981 went to women. By 1996, the corresponding numbers were 44.5 percent of 4,000, the numbers increasingly slightly. In 2000, 44.8 percent of the 5,850 Ph.D.'s awarded in biology went to women. So we're approaching parity.

As for the demographics, both men and women take about seven years to get their Ph.D. in biology and six years to get their degree in chemistry. The median age is approximately the same—32 years for a biology degree and 29 years for a chemistry degree. And the plans to undertake postdocs are basically the same—in biology, 54.4 percent of men and 50.6 percent of women; in chemistry, 49 percent of men and 44.6 percent of women. Although these data are from 1996, the numbers had not changed much as of 2000, except that now almost 70 percent of biologists want to go on and do postdoctoral work and about 50 percent of chemists go on to do postdoctoral work.

Now let's look at the medical school picture. Of the 16,000 students enrolling in 1970, only 9 percent of the class was female. In 2001, 45 percent of the class was female. Just before coming to this meeting I visited my internist, a woman, for a throat culture because I have a cold. After I told her I was speaking at a meeting downtown, she asked me what I was speaking about. "I have to speak to a bunch of medical societies about the role of women in science and how we're progressing," I told her. She then said, "We've solved that problem, right?

When I went to medical school it was 5 percent. Now we're almost half." "That's true," I said, "60 percent of the biologists who are graduating are female, 52 percent of the population is female, but only 45 percent of the medical school class is female. So we've actually got more competition for fewer slots." "Oh," she said, "and we have made progress." I told her that we have, but we have more to do.

Research and Teaching

At the NIH our interests are in research. Recent data from AAMC or from the Federation of American Societies for Experimental Biology [FASEB] show this is a tremendous time to move ahead in translational research. We need well-trained M.D.'s and M.D.-Ph.D.'s who can take science from the bench to the bedside, who can do the translational research needed to take advantage of all the advances we're making right now. It's a very exciting time in science.

Unfortunately, the number of M.D.'s who are going into research is falling. Even though medical scientist training programs are in place, producing well-trained M.D.-Ph.D.'s, we can't compensate for the number of slots we're losing in fellowships and training grants. We need to do something to encourage M.D.'s or M.D.-Ph.D.'s who want to go on and work in medical research careers. In fact, we need to look at this pathway to see how we can get M.D.'s, D.D.S.'s, and D.V.M.'s to move into clinical research. If we don't have these folks, we're not going to be able to take advantage of all the things going on right now. Considering that the growth in degrees in biology and chemistry, both at the B.A. and the Ph.D. level, is among women, the future and growth potential of the societies represented here rests in capturing that expanding portion of the market.

As for what's happening to women at universities and medical colleges, the situation for women on the basic science faculty did not change much between 1990 and 1998 (Table 1). For the medical clinical faculty, the situation is even worse. Overall in 2001, for all faculty, only 12 percent of the full professors at medical schools were women. A slight increase (3.5 percent) is evident at the assistant professor level, and at the associate professor level there has been fairly

TABLE 1 Women Faculty in Basic Science Departments of Medical Schools, 1990 and 1998 (percent)

	1990	1998	Change
Full professor	9.7	13.6	3.9
Associate	19.2	25.4	6.2
Assistant	29.3	32.8	3.5
Instructor	40.9	41.9	1.0

SOURCE: Association of American Medical Colleges.

decent increase (6.2 percent), but if one compares the pool size, it's not good. We're actually losing people. There are places along the pathway where women seem to get "stuck."

A look at total faculty at medical schools reveals that 28 percent are female and 12 percent are full professors. Thirty-six percent of eligible males make the transition from assistant to associate professor, while only 24 percent of eligible females make that transition. So for women, this appears to be the sticking point—from assistant professor to associate professor. Women's attrition rates are slightly higher than those of men, 9.1 compared with 7.7. This then is where we need to focus if we're going to look at that pathway. We have to decide what we need to do to break down the barriers and keep things moving.

Anyone who wants to succeed in biomedical science needs the imprimatur of the NIH in the form of a research grant. In the competition for NIH R-01 support between 1988 and 1997 by new investigators—that is, applying for the first time—there was no difference between men and women on average. They succeeded about 26 percent of the time. So the women are not less well trained, not less competitive in terms of R-01 funding initially.

In terms of new R-01 awards—that is, people applying for a new grant, not necessarily a new investigator—the rates were about the same for men and women, 18 percent and 17.8 percent. As for success rates for renewal of existing awards, again the rates for men and women were similar—35 and 36 percent. So women who get in the system compete equally with men. No better, no worse.

Overall, then, we move students through a system that has more women initially—completing high school degrees, outpacing men in biology B.A.'s, approaching parity in chemistry B.A.'s, but we still have sticking points. We have a lot to do in moving women into the faculty positions where they can act as mentors and role models, where they are competing with the men and moving science forward. Women are 52 percent of the population; they hold 60 percent of bachelor's degrees and 45 percent of Ph.D.'s; they make up 40 percent of instructors and 10 percent of full professors. We can't afford not to take advantage of a labor pool, but all along the way we're losing women.

Some people think in terms of modeling. If in the model of a system there are very small differences in rates but multiple steps, very large differences appear over time. That's essentially what we have here—a system with multiple steps and generally small differences in rates. But over time, when those differences multiply, very big differences occur in the end.

Sixty percent of the B.A.'s in biology are earned by women, but only 20 percent of NIH awards are going to female principal investigators. We must do something about this portion of the pathway in careers. We need to bring more women into the senior ranks.

Advancing Women in Academic Medicine

Janet Bickel, M.A.
Association of American Medical Colleges

The Association of American Medical Colleges, or AAMC, endeavors to link the various women in science and the women in medicine organizations. We maintain a list of about 30 organizations, and our goal is to have all those Web sites linked with ours.

The representation of women in academic medicine from 1977 to 2001 is illustrative of Dr. Carola Eisenberg's talk about "how women saved American medicine" (Figure 3). Imagine our medical school classes and our faculties without the steady increase in the number of women applying to medical school and the number of women physicians becoming full-time faculty at our medical schools.

Because the AAMC does not have much data on medical students or faculty research emphases, I examined our graduation questionnaire, which has been administered to all medical school seniors since 1978. Responses to the only question that asks students to predict their level of research involvement show that interest for both men and women has declined by a third in the last decade (Table 2). This is a big concern for us, that both men and women see themselves as less likely to be significantly involved in research during their careers. The gap between men and women is about the same as it was a decade ago.

The rest of the AAMC data on medical students shows that women are no less interested in science; it's just that they have so many other responsibilities and draws in life. Also, they don't have the mentors and they don't get the one-on-one encouragement aimed at many men to enter science.

The AAMC data on faculty for 2001 show that the distribution of male faculty is fairly even across faculty ranks (Figure 4). Women are more heavily represented in the junior faculty ranks, half as assistant professors. The percentage of women in the instructor rank is over twice that of men.

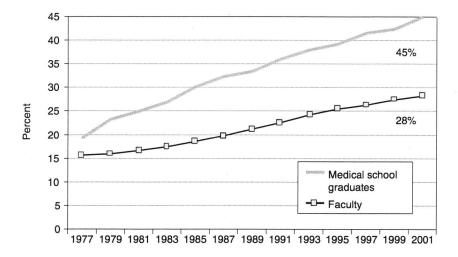

FIGURE 3 Representation of women in academic medicine, 1977-2001.
SOURCE: Association of American Medical Colleges.

TABLE 2 Percentage of Medical School Seniors Expecting to . . .

	1990	1995	2001
. . . be significantly involved in research during their career			
Women	12	9	9
Men	16	14	13
. . . become full-time faculty during their career			
Women	27	—	28
Men	30	—	29

SOURCE: Association of American Medical Colleges.

In 1992 AAMC president Jordan Cohen established a task force on increasing women's leadership. Dr. Diane Wara, who will speak later in this meeting, chaired that task force, and Dr. Page Morahan served as an adviser. The task force looked at four years of school-supplied data, interviews with department chairs, and new research from industry and higher education on women's advancement. Here are the key findings.

• Women make up 14 percent of tenured faculty, 12 percent of full professors, and 8 percent of department chairs.

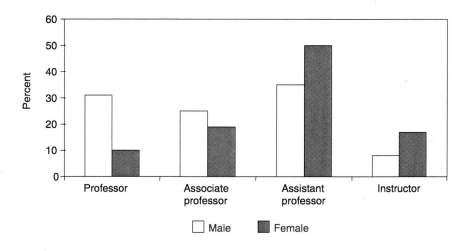

FIGURE 4 Medical college faculty by rank and gender, 2001.
SOURCE: Association of American Medical Colleges.

- Few schools, hospitals, and professional societies have a "critical mass" of women leaders.
- The pool of women from which to recruit academic leaders remains shallow.

Based on these findings, the task force came to the following conclusions:

- The current wastage of the potential of women is of growing importance.
- Only those institutions able to recruit and retain women will maintain the best house staff and faculty.
- The long-term success of academic health centers is inextricably linked to the development of women physicians.

Recent research on women's careers has found that women face many more challenges than men in obtaining career-advancing mentoring. Related to this, many men have difficulty effectively mentoring women. Isolation reduces women's capacity for risk-taking, often translating into a reluctance to pursue professional goals or a protective response such as perfectionism. Finally, without being conscious of their "mental models" of gender, both men and women still tend to devalue women's work and to allow women a narrower band of assertive behavior.

The task force developed the following recommendations:

- *Emphasize faculty diversity in departmental reviews, evaluating department chairs on their development of women faculty.* This recommendation could be carried out by tracking the number of women recruited, promoted, and retained.
- *Target the professional development needs of women within the context of helping all faculty make the most of their faculty appointments, including guidance for men to become more effective mentors of women.* The first step toward implementation might be to compare the costs of faculty development with the costs of faculty turnover.
- *Assess the gender-related effects of institutional practices such as regarding "academic success" as largely an independent act and rewarding unrestricted availability to work (i.e., neglect of personal life).* Implementation might start with creation of an institutional committee to examine practices and policies for their unintended effects on women's advancement.
- *Enhance the effectiveness of search committees to attract women candidates, including assessment of group process and of how candidates' qualifications are defined and evaluated.* One might begin by educating search committees on the pitfalls and opportunities in recruiting women.
- *Lend financial support to institutional Women in Medicine programs and the AAMC Women Liaison Officers and regularly monitor the representation of women in the senior ranks.* A first step might be to conduct a salary equity study and a morale survey.

The most comprehensive analysis conducted to date of initiatives to develop women medical school faculty found that exemplary schools focus on improvements not specific to women: heightening department chairs' focus on faculty development needs, preparing educational materials on promotion and tenure procedures, improving parental leave policies, allowing temporary stops on the tenure probationary clock and a less than full-time interval without permanent penalty, and conducting exit interviews with departing faculty.[8] These schools regularly evaluate their initiatives by comparing the recruitment, retention, and promotion of women and men faculty and by conducting faculty satisfaction and salary equity studies. Surveying faculty about their career development experiences and their perceptions of the environment, comparing the responses of men and women, and presenting the results to faculty and administrators are particularly useful strategies. Such periodic surveys also build a rich data source about why people are leaving and what's keeping the valued faculty at the institution.

The AAMC stands ready to help the societies and the medical schools work to implement these strategies and these recommendations. When all else fails, we

---

[8]AAMC, Enhancing the Environment for Women in Academic Medicine: Resources and Pathways, 1996.

can start making the business case, rather than just the ethical case, for why this work is important. It is likely to mean a better bottom line. Those institutions, according to AAMC President Jordan Cohen, that fail to seize the advantages offered by elevating talented women to positions of power are destined to be eclipsed by those who do.

Participant (Dr. Herbert Pardes): Do you have information on the relative interests of women versus men in going into science in their freshman year or their incoming years before college? I'm looking for what kind of change takes place within the medical schools.

Ms. Bickel: I'll check to see whether the AAMC's Matriculating Student Questionnaire contains that item that I showed you. All we know at this point is that women are more likely than men to lose their research intentions during medical school.

Participant: Do you think your findings are relevant to an individual institution?

Ms. Bickel: Absolutely. We believe the recommendations are just as relevant for individual medical schools as they are for societies. The principles are the same whether adopted by a department chair or a dean or the president of a society.

And like Dr. Morahan, I highly recommend Deborah Myerson's book on tempered radicals and thinking in terms of the challenges and opportunities of being a change agent and when it is possible and when it is not possible to be effective in situations. How do you build coalitions, whether you're the only woman or the only racial minority in a large area? How do you go about making the changes that you believe would benefit the organization and that you particularly value? Obviously, we need the support of meetings like this to inspire each other as well. Isolation is death when it comes to this kind of work.

PANEL

Differences Between Basic and Clinical Disciplines

PANELISTS:
W. SUE SHAFER, PH.D., INSTITUTE FOR QUANTITATIVE BIOMEDICAL RESEARCH
HERBERT PARDES, M.D., NEW YORK PRESBYTERIAN HOSPITAL
JEANNE SINKFORD, D.D.S., PH.D., AMERICAN DENTAL EDUCATION ASSOCIATION

MODERATOR:
SALLY SHAYWITZ, M.D., CHAIR, AXXS STEERING COMMITTEE

DR. SHAYWITZ: In this session, our distinguished panel will discuss the differences between the basic and clinical disciplines. A major goal of this workshop is to increase the representation of women in the leadership ranks of clinical research in American medicine. Procedurally, we hope to extend and adapt the recommendations of this workshop's predecessor, the AXXS '99 workshop that focused on women in the basic sciences. The issues facing women in clinical research differ from those affecting women in basic science research. Our belief is that by better understanding these differences, we'll provide a clearer understanding of the unique issues facing women in clinical research.

So one of our goals today is to delineate those differences, and their impact, and then to determine how to address them. For example, the path for basic science research training is much clearer and shorter than that for clinical research training: college, graduate school, and postdoctoral training—a straight line with no or few professional diversions. For men and women, physicians and dentists, clinical research training is anything but direct. They must contend with four and sometimes more years of medical or dental school where there is little if any focus on research, training, or mentorship or experience in research. Then comes

43

another three to five years or more of residency training—again, typically with meager opportunities or exposure to clinical research.

Just this weekend I spoke to a young woman who is completing her endocrinology fellowship. I asked her what she was planning to do next. She responded that much to her surprise she planned to pursue a career in clinical research. She said, "Through all my years of medical school and residency, no one ever mentioned or spoke of the possibility of a research career. I was never exposed to it, never thought of it. Perhaps if I did, I would have planned better and done things somewhat differently."

So the question is, is her experience typical? Do these perceived differences affect women differentially? How early can clinical societies reach trainees? What can clinical societies do to foster women's interest in clinical research earlier? Can what appear to be structural problems be addressed?

Before going on, let me introduce our distinguished panel. Dr. Sue Shafer is deputy director, Institute for Quantitative Biomedical Research, and former deputy director of the National Institute of General Medical Sciences. Her current interests include biomedical research policy, biomedical ethics, the responsible conduct of research, and enhancing the careers in science of women and minorities.

Dr. Herbert Pardes is president and chief executive officer of New York Presbyterian Hospital and its health care system. Dr. Pardes served as U.S. assistant surgeon general and director of the National Institutes of Mental Health during the Carter and Reagan administrations, and has served as vice president for health sciences at Columbia University and dean of the faculty of medicine of Columbia's College of Physicians and Surgeons. Dr. Pardes has overseen major changes in the education of physicians and enhanced clinical and basic science research. He also has assumed a national role as an advocate for education, health care reimbursement reform, and support of biomedical research.

Finally, but certainly not least, Dr. Jeanne Sinkford is professor and dean emeritus of the Howard University College of Dentistry. Dr. Sinkford has the distinction of being the first woman to serve as chair of a major department in a school of dentistry and was the first African American woman dentist inducted into the USA section of the International College of Dentists. Since 1991, she has been director of the Office of Women and Minority Affairs of the American Association of Dental Schools.

I'm going to ask our panel to consider the differences between basic and clinical research pathways, and their ramifications. There is a difference between how long the pathways take, and how direct the pathways are. There's also a difference in community. Those in graduate training are part of a community of researchers. When those in M.D. training suddenly decide to go into research, they don't have that history of collaboration or community to bring with them. And then there are differences in mentorship and also perhaps in financial encumbrances. So I would like to start by asking our panel to reflect on some of these issues and what they think are the most critical issues differentiating basic and clinical research pathways.

DR. SHAFER: Two images from my recent experience at the University of California–San Francisco draw the contrast between basic research and clinical research in a way that is etched in my mind.

At UCSF, I go to the Program in Biological Sciences seminars held every Friday at noon mainly for the basic sciences faculty. The faculty are supposed to present their work in ways that will promote dialogue, and so they're not supposed to talk about what buffer they used. Rather, they're supposed to talk about the concepts they're trying to deal with. For the last two Fridays one of the more junior faculty members has been at the seminar with her baby. Afterward, half the room sort of surrounds her and talks about how she's doing and admires the baby. That's a picture of the kind of community that surrounds someone who has taken the opportunity, even before she has tenure, to have a child and be very open about that.

By contrast, I went to a talk by Ann Crittenden, who wrote the book called *The Price of Motherhood.* In a discussion after the talk one young clinician was practically in tears. She said, "I'm ready to get out of research. I don't know any woman who has managed to have a clinical research career, see patients, and also have kids. And I want kids." So we immediately put her in touch with a few people like Diane Wara and others to say it is possible. But she was feeling totally isolated and unable to have the kind of life that she wanted, balanced between research and her own life. Those two images for me are the extremes of acceptance and community, and isolation and despair.

DR. SHAYWITZ: Any comments? Our focus here, too, is to see how clinical societies can play a role. Are clinical societies able to intervene at an early stage, helping women who are engaged in clinical research?

DR. SHAFER: Clinical societies have a crucial role to play. In my own home society, the American Society for Cell Biology, women are on par with men in terms of leadership on the council, on program committees, and throughout the society's activities. I know that some clinical societies have started down that pathway. But my observation is that there is much less of that in the clinical world. A society can help to create an atmosphere and take steps to intervene at institutions to place women before search committees and other such things. So I think there is a very strong role for scientific societies in this realm. The trick is to take the lessons we learn today back not just to the choir, the women's committees, but also to the *leadership* of the societies we're trying to influence.

DR. SHAYWITZ: Does anyone in the audience have experience with a clinical society that can serve as a model because it has done just what Dr. Shafer described?

DR. ROBERTA L. HINES (YALE UNIVERSITY SCHOOL OF MEDICINE): I'm not sure I can answer that last question, although again, the Association of Women Surgeons has tried. We do have a little bit of an entrance through the back door, which is the American College of Surgeons. The current executive director of the American College of Surgeons is very friendly, shall we say, and has formally invited the Association of Women Surgeons to participate in programs. So I'm very hopeful that this will be our way into the big boys' organization.

But I do have a question. At my home institution, I've seen some junior women interested in clinical research say, "I can't do this. I'm a clinician; they want me to see more patients and bring in more money, because that's where the money comes from. If I do clinical research I'm taking time away from my clinician duties, and I'm never going to get promoted and tenured." So how can the professional societies help home institutions? How can we get that little credential of having done clinical research, perhaps sponsored by one of the professional societies, to counteract the negativity of less clinical revenue at the home institution? Is there a way we can get the professional societies to interact with universities in that way?

DR. PARDES: I think you've put your finger on a central point. The problem of adequate numbers of clinical researchers has been well documented, and the very points just made are ones heard from *both* women and men. The problem is, when one adds to that set of problems surrounding the clinical research career itself the other kinds of problems that are obstacles to women moving ahead in academic medicine, the difficulties are compounded. In other words, two sets of problems exacerbate one another.

Now let me preempt my own comments. I was a member of a clinical research panel that Harold Varmus, former director of the National Institutes of Health, convened and David Nathan, professor of pediatrics at Harvard Medical School, chaired a few years back. The NIH has made some very good moves in trying to respond to some of the needs of clinical research. I argued at the time and I continue to argue that the clinical research panel should not have been disbanded, because this kind of issue is an ideal one for an NIH committee. I'm familiar with all the other advocacy groups, but such groups don't cut the same way as one that's perceived to be advising the NIH on how to proceed. The question of how to facilitate the movement of greater numbers of women into successful clinical research careers is a legitimate enough issue to ask the NIH to reconstitute its panel and put that as one of the priorities on the panel's agenda.

Now I want to segue into another point: the entire question of university attitudes toward people making accomplishments in clinical research. I am a faculty member at two universities, and at one, at least, the tendency to see clinical research as less valuable than basic research is compounded. One of the biggest problems is some of the faculty themselves.

One of the thoughts we've had about how to solve that problem—"we" meaning the clinical research forum headed by Bill Crowley at the Association of Academic Health Centers—is to encourage the NIH leadership to come together with university leadership, by which I mean university presidents and deans, so that the source of funds for research, the people who control those funds, talk to the leadership of academic institutions about this problem.

DR. MICHAEL LOCKSHIN (WEILL COLLEGE OF MEDICINE, CORNELL UNIVERSITY): For the past decade I've been on various committees about clinical research, both at the NIH and outside the NIH. One of the issues that should be on the table is the very broad definition of who goes into clinical research. It encompasses the people who take blood or other specimens from individual patients and basically never get out of a laboratory, all the way up to the people who participate in drug trials, and those sorts of things, or outcome studies.

I would posit, and possibly in an inflammatory way, that the issue is not so much one of gender. It's that the larger-scale clinical activities are always ones that require large amounts of collaboration, result in multiple names on papers, and almost never end up in the individual being the first author on one thing or another. Such activities often take a very long time to accomplish and result in one or two large-scale papers as opposed to items in the journals that publish very rapidly and primarily in the basic sciences.

If there's a positive way to look at this and to put it in a gender context, and if women are in fact the better collaborators and the better sharers of information, then putting value on collaborative research, the multidisciplinary or multi-institutional types of research, and making that value their own would be a way to do that. As Dr. Pardes brought up, to be worthwhile, that value must be recognized at the institution level by promoting people for that type of activity as well.

To summarize, the issue is to define what is meant by clinical research when you're trying to describe whether or not women are advancing in that area. Then you need to reward those components that do not necessarily lend themselves to the same measurement criteria used for promotion in basic science activities.

DR. SINKFORD: I'd like to speak to several issues that have already been mentioned. Janet Bickel spoke about women in medicine, and most of the things occurring in dentistry are parallel. Female enrollment in dental schools has gone from 2 percent in 1970 to 40 percent in 2002. Twenty-five percent of the dental faculty are now women. If we were to lose those 25 percent, with 200 vacant faculty funded positions available in dental schools, we would really be in trouble. So we see women as a very critical resource for the future development of faculty, research, and community programs.

Our advanced programs for dentistry are similar to those in medicine, except that there is no required residency program. So about 36 percent of our students go into advanced training programs in the specialty areas such as orthodontics, periodontics, and prosthodontics. Those programs usually include some research requirements, but they're not extensive. As a result, men and women graduating from those programs are specialty qualified, but they are not clinical research qualified. That's where we have a gap in our ability to take those individuals who have completed advanced training programs and expect them to do clinical research. They just do not have the skills to do that kind of research without preceptors. I think similar things occur in medicine.

We've asked the National Institute of Dental and Craniofacial Research to help us secure some bridging grants so that individuals just leaving training programs and joining faculties or hospital staffs could have preceptors and therefore some kind of linkage with a research substructure that enables them to perform. We need to find ways of continuing that training and that mentorship. Those individuals do not have grantsmanship skills, which are very important. If they're going to get money to support their research, then they must find a way to develop those skills or be able to call on the infrastructure within their institution to help them write appealing and competitive grant applications.

At a summit held last year we brought together all the 55 dental school deans from across the country, with their chancellors and their presidents, to talk about enhanced clinical research within their institutions across disciplines and within the health science centers. We're trying to have an impact on how the research capacity of our schools will develop over the next few years. That capacity doesn't affect just women, but also male junior faculty.

This meeting is a way for academic institutions to partner with our societies. Much of clinical research could be undertaken through collaborative ventures. But we have not pursued that vigorously, either in medicine or in dentistry.

DR. SHAFER: I just had a comment on your discussion, Michael [Lochskin]. I see a role for scientific societies to find ways to reward junior faculty in a discipline in which the norm is large collaborative projects. Each society could consider taking a strategic look at its particular discipline and seeing how it can make recognitions externally that can be used internally.

DR. LOCKSHIN: That is a good idea. I think it is appropriate to bring up and value the collaborative components. Within even large-scale projects, individuals have original contributions. I think it would be a wonderful idea for a society to give a prize or additional money to a woman who has distinguished herself in that capacity. That would work very well within the usual criteria for promotion-national recognition and that sort of thing.

LYNN GERBER, M.D. (NATIONAL INSTITUTES OF HEALTH): To pick up on something Dr. Lockshin raised, I work at the NIH, but I'm representing the American Academy of Physical Medicine and Rehabilitation. It's a very practical group of clinicians who are quite ambivalent at some level about whether a science is associated with rehabilitation. It's a struggle that's going on right now fairly vigorously within the academy. We've had lots and lots of recognition for clinicians and teachers and all kinds of people who have made enormous clinical contributions to our field, but we're very shy on the research side. In fact, a lot of discussion is under way right now on how to pull up the research activities of the academy.

With that in mind, the education and research fund has begun to target women, so that, interestingly enough, various kinds of scholarships and travel funds are available to encourage women to participate in research activities, attend conferences, and make presentations. It's a little early to know whether this effort will reach fruition, but what has emerged, as a result of women being a little bit more prominent in our academy, is the notion that maybe our academic centers should be looking at ways of providing tenure that do not rely so heavily on research. We are not a heavy research organization. For example, could we give tenure to people for very strong track records in teaching in clinical practice? As for what clinical research is all about, it needs the clinic. Often women are the clinicians who are providing the excellent care and the excellent research information that is the fodder for the statisticians and basic researchers who come up with the outcomes of those trials.

So the academy is now looking at whether we can lean a little bit less heavily on research and developing track records as first authors and look more toward the three legs of the medical school and the academic establishment, which include clinical practice and teaching.

DR. KAREN ANTMAN (COLUMBIA PRESBYTERIAN MEDICAL CENTER): At least in cancer, the clinical research track is fairly well defined. A junior faculty member collaborates with a laboratory at the cancer center or university. She has an idea that she then takes to the clinic and does a Phase I trial. If the Phase I trial shows that the idea is safe, she then does a Phase II trial. If she happens to be the investigator that has a Phase II trial that looks interesting, she goes into a cooperative group and does a big Phase III trial. She gets the leadership role in that trial, and so is now first author of a national paper. Her institution is not supposed to be putting patients into the Phase III trial because she's now supposed to be back doing the Phase I and Phase II pilots for the Phase III trial at a national level.

So this process allows clinical researchers to actually take leadership roles that are well defined for promotion committees and to achieve first-author publications—not in *Science* and *Nature*, but certainly in the *New England Journal of Medicine* and the *Journal of the American Medical Association*. It's hard to put this track in place where the culture's not there, but the track is well defined.

DR. SHAYWITZ: Do you think more women are represented in that track?

DR. ANTMAN: At least in cancer, plenty of women are represented in those doing the Phase I and Phase II trials and then, if the data look interesting, moving on to a Phase III trial and being the first author.

DR. DIANE WARA (UNIVERSITY OF CALIFORNIA–SAN FRANCISCO): Just to expand on Dr. Antman's statement, the National Cancer Institute–sponsored/funded cancer centers have insisted on door-to-door, back-to-back clinical investigations and laboratory-based investigations—that is, they have insisted on collaboration between the two. So I give the NIH some credit for the development of, I believe, the largest body of clinical investigators in the country in cancer.

DR. ANTMAN: But sometimes the laboratory investigators don't buy into this model, even though it is the model that is funded by the National Cancer Institute [NCI]. So we're caught between the culture of the university and the culture of the NCI, right in the middle.

DR. WARA: I wanted to make a different comment, however. The NIH funds 81 clinical research centers across the country. These clinical research centers have an annual meeting and a society. For the last decade, we have brought funded junior investigators to this meeting. About three years ago, the clinical research centers brought in the K-23 and K-24 award recipients (K-23s are mentored junior investigator training grants, and so they are in the name of the recipient, the mentee; K-24s are more senior). This year, we're going to include the K-12s, those eight who may have been funded, and we are inviting every funded K-23 investigator to the meeting to present his or her research. The Association for Patient-Oriented Research [APOR], led this year by Dr. Leon Rosenberg of Princeton University, will be coming to this meeting as well, as in past years.

I think what we've heard today is the importance of intervening at a fairly junior level in order to nurture and expand the body of clinical investigators in this country. The meeting I described is an example of the NIH collaborating with societies, because both of the societies mentioned are NIH-based groups. It also is an example of focusing on junior investigators to ensure that we engage them early in their careers in clinical investigation. This may be the largest group of clinical investigators in the country, and we should try to engage that group. The group badly needs help, and we desperately need help in terms of a model so that we can help them.

DR. PAGE MORAHAN (NATIONAL CENTER FOR LEADERSHIP IN ACADEMIC MEDICINE): I also want to amplify the talk on how to broaden the view of scholarship. It's critically important, because scholarship has hierarchies. Applied scholarship is lower on the totem pole, as is interdisciplinary scholarship. Certainly our universities can provide one approach to broadening the view of scholarship; they could broaden their own internal rules for promotion and valuing broader scholarship. But that's not enough, because in this day and age when faculty move from one school to another they need to know that they have the "scholarship union card" they can take with them to other universities.

So I'm very much pleased to hear about the collaboration of societies and the efforts of societies to broaden their own views of scholarship. That's where we need more activity. This has been done to some extent in the basic sciences, such as the American Chemical Society and others, where the American Association of Higher Education undertook a granting process to help each society develop a broader view of scholarship. We need to do the same thing through the Council of Academic Societies.

NANCY SUNG (BURROUGHS WELLCOME FUND): I was very excited that Dr. Wara mentioned the clinical research meeting, and I would like to add a few more things about that meeting, perhaps as fodder for discussions at this meeting. A portion of that meeting described by Dr. Wara was sponsored by five foundations working in collaboration: the Burroughs Wellcome Fund, the Doris Duke Charitable Foundation, Howard Hughes, Robert Wood Johnson, and Juvenile Diabetes.

Each foundation was receiving funding requests from multiple clinical research societies for duplicative activities. We felt that a collaborative effort might actually reach more people and be more substantial. So the clinical research meeting Dr. Wara referred to actually had a career development track in it that provided a mock study section—a session on the new NIH loan repayment program as well as networking opportunities. We discovered that many of the trainees attending those meetings were also part of smaller specialty societies that did not have a full-blown career development track for their young investigators. We believed that if the meeting could grow to include perhaps some of the smaller societies that don't have the capacity or the funds to mount such an effort, it could act as a resource for these people similar to that provided by the Federation of American Societies for Experimental Biology for basic scientists. These funders are actively thinking about this, and they would love to hear your thoughts on it.

JANINE SMITH (NATIONAL EYE INSTITUTE): Long ago there was a question about how individual societies or groups might be able to effect change within their groups. The American Academy of Ophthalmology formed a subgroup called Women in Ophthalmology, which gave women one way to get a specific forum at the academy's national meeting and to invite speakers that covered topics from women's health issues to things such as recognizing retinal changes associated with battering. So one approach is to form a group within your academy and then seek to place a member of that group on the board—our goal in the future.

I also would like to comment on Dr. Lockshin's and Dr. Antman's comments on clinical research and what it is. This issue is different for every specialty. It is quite straightforward to develop an anticancer therapy—Phase I, Phase II, Phase III—and then seek the approval of the Food and Drug Administration. The outcome measures for those trials are quite clear; it's mortality. For other fields, it is not that straightforward. For example, natural history studies—the epidemiologic studies needed to get prevalence data in order to do valid sample size calculations for those clinical trials—have to be done. Those are not as straight-forward and take a very long time.

Clinical research includes many types of studies, and I don't think we all recognize that. In fact, there is a Society for Clinical Trials for people who are clinical trialists. That's another meeting and another society that we should add to our list.

ESA WASHINGTON (JOHNS HOPKINS UNIVERSITY, SOCIETY OF TEACHERS OF FAMILY MEDICINE): I just completed a fellowship, the Robert Wood Johnson Clinical Scholars Program. Many of my colleagues and I are concerned about the debt issue, which is very big. A woman comes out of residency with a $100,000 debt at a minimum, if not more, and she can't defer in residency like she could have done several years ago. Does she go into a fellowship for another two to three years, at a salary of $30,000–40,000 plus minimal benefits? This situation is even more difficult when children are added to the mix.

We need to figure out some way to give fellows and residents financial planning advice along with the career development advice so that they can some-how balance their situation and really believe that the short-term sacrifice they're making will pay off in the long term. It's one of the things that's extremely challenging for fellows and extremely stressful. They're trying to think of a research question and do a great project and get the papers published, all while trying to deal with large personal financial responsibilities with little support.

DR. SHAYWITZ: I'm so glad you brought that up, because that's such a prevalent issue and one that people don't speak about. That will give us something to talk about during our breakout sessions and perhaps as we sum up.

DR. ANTMAN: That's why mentors are necessary. And to comment elsewhere, certainly clinical trials are only a small part of clinical research. I usually like to describe the continuum of research as laboratory, clinical, and public health. Public health has a whole different aspect of research. Population-based research is quite different from clinical research and quite different from laboratory research. Good health research cancer centers and medical centers need a con-tinuum of all three, I believe.

JOAN AMATNIEK (JANSSEN PHARMACEUTICALS): I do clinical research and I don't work in an academic center; I work for a pharmaceutical company. I'm only three years out of a fellowship. The opportunities for women to do clinical research in particular in pharmaceutical companies cannot be beat by universities anywhere. My boss doesn't just compete with the competitors for me, he competes with my family for me and he knows it. So in my work life, I get to do great research, I get to work with experts, I get to design those big trials that other people work on, plus I work for someone who doesn't want me to abandon my family.

I would like us to think about where clinical research is done. It's not done just at the university. Think about other options and perhaps bigger strategies. Perhaps in the early part of a woman's career, after her fellowship, it's better that she work in industry and be able to have these leadership opportunities. Later on, when her kids are bigger, she could bring those skills back to the university.

DR. PARDES: Your point's very well taken. Sometimes we tend to think in terms of there being only one place to do research—universities. But having said that, I think that I would not want that to divert our attention from the fact that within the university culture itself there's still a problem we have to deal with.

DR. SHAYWITZ: These wonderful comments have broadened our purview, both of how much progress has been made and of the problems that still exist. Now I'd like to ask our panel to sum up their perceptions of one or two key points that they would like to leave with the audience to serve as a basis for our discussions in the next part of the workshop.

DR. SINKFORD: I was struck by the comment about funding. I think some programs are now in place to help with loan repayment. We should follow up on that as a means of keeping the junior researchers in the pool and making it easy for them to make a decision about a long-term career.

DR. SHAFER: The theme I heard is not on any individual point but on the need for us, as we think about what problems we're trying to solve, to be strategic. Take a very specific problem and try to take a strategic look at it.

DR. PARDES: If I could make two or three points. First, I want to go back to Dr. Marino's data, because I think they were very instructive. She told us that about 45 percent of medical school students are women. But her 1998 figures show the percentages going down as women move up the ladder: from 41.9 percent as instructors, to 32.8 percent as assistant professors, 25.4 percent as associate professors, and 13.6 percent as full professors—substantial drops in each of the successive faculty slots. So something is happening in a continuing way and a look at it would help us focus on the problems.

Second, I think mentorship should not just be mentioned, but should be the subject of a full-focus discussion unto itself. What is it? How does it get done? Why do people do it? How do we give people the incentive to do it? I was happy that at the NIH some of the new grant mechanisms they established in connection with the clinical research thrust included some faculty support for mentoring. Mentoring should be a criterion perhaps for promotion, or for tenure, or for financial incentives. I think that would help. Mentoring is a complicated business, yet most people in research careers would acknowledge that the quality of mentorship and the kind of leaders one can turn to—both immediate, ongoing

mentors on-site and then contacts around the country through societies and such—
can make all the difference in whether a person feels nurtured and is able to
blossom into a real scientist.

Finally, when we ask whether we should turn to societies or universities in
dealing with the problems, I suggest that we look for some way to convene the
leadership of several entities—the NIH, the universities, the societies, the pharmaceutical companies—to say, all right, there's something that each one of us
could do. Is there some way in which, by collaborating, we could create a fabric
and have on each of our agendas people whose reward, whose gratification in life,
is to address this issue and promote it? That is, they are put on the spot, if you
will, for seeing that these kinds of figures move in more dramatic ways throughout the discipline.

Women in Leadership

Ruth Kirschstein, M.D.
National Institutes of Health

Speaker Introduction
Sally Shaywitz, M.D., Chair, AXXS Steering Committee

We have talked a lot about the role of National Institutes of Health and how much the NIH has accomplished, and so it is only fitting that our next speaker is a wonderful representative of the NIH, Dr. Ruth Kirschstein, deputy director of the NIH and until recently its acting director. Dr. Kirschstein has served as director of the National Institute of General Medical Sciences and as acting associate director of the NIH for research on women's health. She has received many honors and awards, most recently the Albert B. Sabin Heroes of Science Award from the Americans for Medical Progress Education Foundation. She was also recognized by the Anti-Defamation League, which bestowed upon her the Woman of Achievement Award. We're honored to welcome her and hear her comments.

Dr. Kirschstein:

I want to thank you so much for inviting me to address this workshop. The excellent report on the first workshop, AXXS '99, serves as a base to follow up on, as does the work of all the people serving on the planning groups and the many planning activities that have occurred over many years. Indeed, as one reviews the work of these many groups, it should be noted that all of them are made up of people who have been doing this for a very long time.

As I was thinking through what I might say to you, I did something over the long holiday weekend that I haven't had much time to do for quite some time— I indulged myself by reading a weekday, albeit Independence Day, issue of the *New York Times*. Lo and behold, there I found the theme for my talk today, on the first page. The headline read "Many Women Taking Leadership Roles at Colleges." My friend and colleague Shirley Tilghman was featured in this article. The *Times* said that in the year since she has assumed the role of president of Princeton University, the previously male-dominated university and an Ivy League school, she has appointed four women to top administrative posts, and she has retained and reappointed one more who had been appointed by her predecessor.

In just a tad more than the 30 years since Princeton opened its doors to women for the first time, says the *Times*, "the changes at Princeton are a signal moment, or an occasion to take stock of the fundamental shift but also to think about how much more is left to be done, in terms of senior faculty and administrative positions."

For years, those of us in leadership positions who were women had always assumed—and we still do—that mentoring and nurturing were essential to the development of careers for women in academia, government, or even business. We believed that for women to move up the chain it was essential that they move along in such a way that they could assume positions of leadership, accruing more and more power. In other words, we believed that women in higher positions followed an orderly path of advancement. Indeed, that is usually what happens.

But Shirley Tilghman points out that in these times we perhaps need another course, a concerted effort to *find* women for the top positions. As the *Times* quoted Shirley, "Twenty-two percent of the faculty but only 14 percent of the faculty of women are full professors." The key was to appoint more women administrators to build on the women who are presidents of key universities and colleges.

It's with immense pride that we can now say that we have 11 women presidents at major universities and colleges. That was the subject of an article published in *Newsweek* magazine, just four days before the *New York Times* article. Just four days, indeed, before Independence Day. What a wonderful way for women to celebrate their independence. These women are Hannah Gray, Jill Conway, Nan Keohane, Donna Shalala, Jonetta Cole, Ruth Simmons, Judy Rodin, Shirley Jackson, Mary Sue Coleman, Nancy Cantor, and Shirley Tilghman herself. All have a network and all have had, in one way or another, a hand in mentoring others. In addition, they frequently talk to each other. They have learned to cover the full spectrum, from top administrators all the way down to burgeoning faculty.

They start with the assumption that too few women hold high-level faculty positions and therefore are unable, supposedly, to see that there are more goals they can attain. Some of these women could be appointed to those high levels early and be taken a chance on, so to speak, because they undoubtedly will be able to perform. Men have known that for a very long time. There are many examples of individuals who have been chosen for a major position, somebody who did not necessarily move up the chain.

We as women need to maintain our ability to go out there and find the talent. We need to keep people in mind for these kinds of major positions.

We also need to do something about the major problem, as you've heard all through this workshop, of what we are going to do about child care in the United States. Other countries have solved this problem. We so far have not been able to do so. But all of us working together must forge the answer to this very vexing problem.

Finally, we must strive to stop placing the modifier in front of the noun. To stop referring to a "woman" president or a "woman" dean; she is a president or she is a dean. If we do that, I think we will set the stage for some of the things we really want to see happen. With that and through what I hope will be a continued set of workshops, we will build on success that modifies itself constantly for more and more women in those top positions and others as well.

SESSION III

Reports of
Breakout Sessions

Reports of Breakout Sessions

After the presentations, workshop participants attended breakout sessions to elicit suggestions that might be taken back to the societies represented. Groups were asked to try to identify the specific obstacles that prevent women from advancing in clinical research and then to determine what could be done to address and overcome these obstacles.

Some of the topics for discussion were leadership, visibility, and recognition; mentoring and networking; best practices (e.g., how can societies implement a five-year plan to ensure that leadership reflects each society's demographics); and oversight, tracking, and accountability. Breakout group members were encouraged to ask themselves what specific steps societies can take to ensure that more women assume leadership roles, and how societies can be convinced that diversity is in their own best interest, that it is critical to their mission? What successful approaches, model practices, and programs have worked? How can they be adopted by other societies? What would an ideal program look like, and what would it take to make it happen?

Leaders of the breakout sessions then presented summaries of the suggestions identified in each of the groups. These suggestions reflect the views of the individual presenters and do not necessarily represent the views of workshop participants as a whole.

NANCY ANDREWS, M.D., PH.D. (HARVARD MEDICAL SCHOOL)
LEADER, BREAKOUT SESSION 1

This group took a "strategic" approach to its mandate:

1. Think strategically.
* *Capitalize on the current culture and mindset*: remind those in leadership positions that they will be missing opportunities to enrich their own societies, their own institutions, if they do not capitalize on the women in science.
* *Take advantage of the current culture of networking*: it's not just the old boys' network anymore.
* *Change the current culture and mindset:* (1) look at the definition of academic success in the appointment, promotion, and tenure process; (2) value mentoring for what it is and recognize it as a very important part of academic success; (3) look at the definition of scholarship, emphasizing and developing better metrics to incorporate and reward those who engage in collaborative and clinical research.

2. Act strategically.
* *Collect better data* on clinical researchers. A survey might be undertaken with the lead of AXXS to determine the demographics of societies and whether the leadership and staff of those societies reflect their memberships.
* *Look at some of the other equity issues* such as salary, perhaps starting with the American Association of Medical Colleges (AAMC), which does comprehensive salary surveys each year.
* *Collect hard data on recruitment versus retention costs* to suggest just how cost saving retention is. Related to this, promote a strategy of internal recruitment of women as well as external recruitment of women.
* Once these data are available, *disseminate the data*, with the help of the societies, to department chairs and the society memberships.

3. Search for models, institutional-based and society-based, in three areas:
* *Career development,* which covers the categories of financial, academic, and scholarship. Specific areas might include grantsmanship, conflict management issues, negotiating skills, full career development workshops.
* *Mentoring:* determine awards and rewards societies are developing in this area so these models can be shared with others: The American Society of Hematology awards big grants for the mentee in which some of the evaluation criteria are related to the mentoring skills involved. Now the society is working toward rewarding the mentor financially.
* *Recruiting and advancement:* awards from national societies might be used to award department chairs or other leaders for appropriately recruiting and advancing women in science.

W. Sue Shafer, Ph.D. (Institute for Quantitative Biomedical Research) Leader, Breakout Session 2

This group identified four action items that it thought could be implemented through the professional societies:

1. Facilitate and highlight the value of *mentorship* through mentorship awards sponsored by the national societies. Look for ways to institutionalize the accountability and value of mentorship.

2. Provide a mechanism for *ongoing interaction between mid-level and senior-level women* through society meetings and by societies working together to share information.

3. Encourage *editorial boards of societies* to ensure that their boards reflect the demographics of their memberships. Governing bodies and elected officers should likewise reflect the societies' makeup.

4. Promote *collaboration and interaction among societies*. They should share information, avoid duplication, highlight well-functioning models, and continue the conversation begun at this workshop.

Michael Lockshin, M.D. (Weill College of Medicine of Cornell University) Leader, Breakout Session 3

Our group started with rather small problems and then moved into the larger issues. It made the following suggestions:

1. Analyze the *infrastructure issues* that affect women and men in the workplace: flexible work schedules, social support for child rearing and child care, administrative support, space issues. Recognize that local institutions (such as universities, hospitals, and research centers), societies (such as AAMC), and national institutions (such as the National Institutes of Health) can all put forth the arguments required to instill the flexibility needed in the system.

2. Promote *mentoring* at all career levels, from predoctoral on up, that can be supported by individual institutions, societies, and national organizations.

3. Seek *role models* of successful women who have managed to balance families and successful careers and make them highly visible. By constantly seeking new role models, one can avoid overburdening a few people.

4. Publicize *fiscal models* that support retention of staff, comparing the cost of attrition with that of training new staff. Specific programs include debt forgiveness, staff reentry programs, and flexible arrangements.

5. Present these issues as cutting across disciplines to funding agencies and Congress.

HERBERT PARDES, M.D. (NEW YORK-PRESBYTERIAN HOSPITAL)
LEADER, BREAKOUT SESSION 4

This group identified five areas that require emphasis.

1. Ways in which women can more successfully balance their *personal lives* (including elder care and child care) *and professional development.*

2. The *criteria for promotion and tenure* in universities. Organize a consensus conference on such criteria?

3. M*entoring*, from defining it to finding incentives for its adoption. Mentors must feel valued, and so links should be established with the criteria for promotion and tenure. The following concerns were raised about mentoring:
   • Within research institutions, it is better that mentoring relationships be established between the mentee and more than one senior person (mentor)—for example, a mentee might have both a personal and a scientific mentor.
   • Mentoring raised concerns because established clinical researchers seem to live more fragile lives than some of the basic scientists; placing young people in mentoring relationships with those whose own lives are precarious may present problems. In a related area, basic scientists may find it easier to extricate themselves from the mentor's lab and establish their own independence than clinical researchers, who are so highly entangled in clinical investigations and collaboration.

4. What is special and different about the *clinical research environment.*
   • Young men and women from a medical training environment have far less in the way of basic training in research methodology, statistics, and many other areas that allow a researcher to succeed. As a result, more programs based on the curriculum programs run by the NIH are needed, and stipends should be attached.
   • As for the role of the NIH, its debt forgiveness programs and financial support for mentors are laudable. It is also paying greater attention to the review of clinical research to ensure that clinical research gets a fair shot in its reviews rather than being overseen by a review panel made up of 11 basic scientists and a token clinical researcher. But the NIH should be in continuing contact with its consumers, its grantees, to see how it can better foster careers, and it should

reconvene a clinical research panel similar to the one chaired by NIH Director Harold Varmus five to six years ago, because that kind of panel becomes a continuing prod and kind of overseer to ensure some of these things happen.

5. *Societies* as agents for the development and advance of young clinical researchers. They can provide courses in how to negotiate and how to request resources and support from a university. If societies see themselves as allied with new scientists coming into clinical research, and as agents who will help these scientists succeed, they may often be able to do things that universities cannot do in this area.

JEANNE SINKFORD, D.D.S., PH.D. (AMERICAN DENTAL EDUCATION ASSOCIATION) LEADER, BREAKOUT SESSION 5

This group addressed three questions: What obstacles do societies face? What effective practices are already in place? What would we suggest as the outcome of this workshop?

1. Obstacles
• Clinical research is usually viewed as a second-class citizen within institutions and societies, and in many contexts, not the least of which are promotion and tenure, it does not have the same impact as other areas. Therefore, efforts to encourage people to undertake this kind of research will not fly when there's no reward for it.
• Financial issues and lifestyle issues are obstacles as well. Clinical researchers are often not well funded, and many have loans to repay. They need better salary and better revenue. As for lifestyle concerns, they are pinched from every direction—clinical research, service activities, family care, and myriad other demands. Everything more they do in their professional life takes away from family or personal life.
• A lot of women do not feel that they *are* role models. That perception needs to be changed, because everyone is a role model regardless of her level.
• Lack of access to information channels and grantsmanship channels is a barrier.

2. Effective practices
• The American Academy of Family Physicians (AAFP) has programs already in place to reach out to medical students. A small minority of those students become family practitioners, but they have had access to some clinical research programs. An AAFP committee on special constituencies looks at the needs of different groups.
• The Society for Teachers of Family Medicine at its national meeting

encourages resident and student research presentations that then are used as a basis for discussion at the main sessions of the meeting. These students or residents are paired with senior mentors before the meeting, and the latter guide the young people through the meeting and encourage them to get involved in clinical research.

• The National Institutes of Health has a very successful loan repayment program.

• The Association of Women Surgeons (AWS) and its tax-exempt foundation have several successful programs that include a clinical research fellowship in minimally invasive surgery funded by Ethocon Endosurgery, a visiting professor program for women, its "Pocket Mentor," and an online mentoring program.

3. Action items

• To move clinical research out of the role of second-class citizen, define criteria for excellence in clinical research, present them in some venue such as a meeting of the Council of Deans of Medical Schools, and seek endorsement of the criteria by the academic health centers so that they can be used as valid criteria for promotion and tenure. The societies would have to play a very significant role in such an undertaking, because each specialty would have to develop its own criteria of excellence in their area.

• Encourage each society, with the help of the talent present at the AXXS workshop, to put on leadership programs in the context of the national society meetings or research training programs.

• Encourage the individual professional societies to collaborate with the NIH and the private sector for joint funding. When the funding from a K award (NIH training) grants is not enough, make it allowable for the professional society to add its particular grant and not exclude people because they already have another form of funding.

• Pursue the possibility that through AXXS a pool of funding could be established for the initiatives just outlined.

SESSION IV

Closing Plenary

Achieving XXcellence

Diane Wara, M.D.
University of California–San Francisco

SPEAKER INTRODUCTION
SALLY SHAYWITZ, M.D., CHAIR, AXXS STEERING COMMITTEE

Dr. Wara is associate dean for minority and women's affairs at the University of California–San Francisco. As a key member of the chancellor's advisory committee on the status of women, she guided the passage of a number of faculty changes, including the statewide University of California policy on child-bearing and child-rearing leave. Dr. Wara is division chief of pediatric immunology and rheumatology and program director of the pediatric clinical research center. She is also an expert on abnormalities of the immune system in children, has a primary interest in AIDS, and has published extensively in this area.

DR. WARA:

I wear two hats here. I've just completed 12 years as the associate dean for women and minority affairs at UCSF, with fairly heavy involvement at the Association of American Medical Colleges [AAMC] during that same time period. As well, I'm a clinical investigator. I tried to put those two hats together in fashioning a fairly brief summary talk that will contain these four components: What is clinical investigation, because that is what we're here to talk about? Are women faculty in schools of medicine advancing? I will then offer a personal perspective. I finish by asking how can professional societies enhance the development of careers for women in clinical research?

Clinical Investigation—What Is It?

What is clinical investigation? If one takes just the middle component of Karen Antman's notion about what do we do at academic institutions—the first being laboratory-based investigation, the second clinical investigation, and the third outcomes-based and epidemiological research—then clinical investigation in the middle implies that a doctor is in the same room with the patient. That definition was put forward by three different groups, including the Nathan panel that met four or five years ago at the National Institutes of Health [NIH]. By the way, having the doctor in the same room as the patient has some fairly heavy implications for the investigator.

Over the last decade the faculty in clinical investigation has decreased by about two-thirds—not the senior people who were well established, but the junior people coming in and moving up as clinical investigators. The problem, if you will, begins during medical school when clinical investigation is not discussed very thoroughly. It then moves on into residency where we really don't spend a lot of time with them talking about clinical investigation. Those young residents in decreasing numbers have been choosing to enter subspecialty fellowship programs. During that time, a few are snared; we do glean some clinical investigators through our fellowship programs. But there's been a significant falloff.

The data come from three separate reports during the 1990s.[1] The first was published by the Institute of Medicine in 1994. The second is the Williams report published in 1995 (I was a member of the Williams panel). Gordon Williams and some of my colleagues in this room from the National Institutes of Heath, as well as some of us from academic institutions, looked at the grant funding at the NIH. We asked ourselves what was good and was directed toward clinical investigation, was there enough, and was it in the right proportion for different issues. And we made suggestions, most of which were followed. Finally, there was the Nathan report, published in 1998, which advanced both the IOM and the Williams report and put forth solid recommendations to the NIH.

The data from these three groups were consistent across the groups. So what's happened? Since the Nathan report, the dollars at the NIH have increased by about one-third for clinical investigation. That includes all the mentor career development awards that we've heard about today, which really are aiming at the crux of the problem: how to bring more young investigators, both men and women, into clinical investigation. The answer is to provide structured mentored

[1]Institute of Medicine, *Careers in Clinical Research: Obstacles and Opportunities* (Washington, D.C.: National Academy Press, 1994); G. H. Williams, Chair, "An Analysis of the Review of Patient-Oriented Research (POR) Grant Applications by the DRG, NIH Clinical Research Study Group, " National Institutes of Health, Bethesda, Md., 1994; D. G. Nathan, Chair, "The NIH Director's Panel on Clinical Research Report to the Advisory Committee to the NIH Director, December, 1997," National Institutes of Health, Bethesda, Md., 1998.

career development awards as bridge dollars at the end of a young person's fellowship or at the beginning of her academic career, and before she should be expected to compete for R-01 grant support. Ideally, there should be about five years of bridge dollars.

Study sections actually have been changed, although quite modestly. Some study sections have had the Gordon Williams golden rule imposed. I had the great job of reviewing for one calendar year the abstracts from every NIH grant submission if either "I" or "B" or clinical investigation was checked on the front sheet. When we looked at those abstracts and then asked how many of these grants were funded versus the priority scores or ranking of those grants by a study section, we learned (and we published in the *Journal of the American Medical Association*) that grant applications were much more likely to have a good and fundable priority score if they were reviewed by a study section that included at least one-third peers. For clinical investigation that meant that one-third of the study section should be or have been intimately involved with clinical investigation.

The NIH took that finding to heart and changed about 20 percent of its study sections. The result is that grant applications that move through the modified study sections, which have a larger number of clinical investigators, do better. In other words, applicants for clinical investigation do better if they're reviewed by their peers. That's simple.

The difficult part has been figuring out how one coerces enough faculty from our academic institutions to fill out one-third of the study section slots in the appropriate study sections. So I recommend that societies strongly encourage their members to participate actively in study sections. When asked, don't say no, because it's really where our dollars, our future comes from.

So study sections were changed, and that was good. However, women in clinical investigation and their unique needs were not truly addressed. Men and women have the same startup issues, and they have the same mentoring issues. But there is a difference between a young woman and a young man entering clinical investigation in academic medicine. Two of the major issues have been mentioned. The first is that whether we like it or not, agree with it or not, raising our children and caring for our parents remain primarily a woman's job. Therefore, the impacts of those two care-providing jobs—I would say privileges actually, rather than jobs—fall on women. How to make that work for a young, mid-level woman clinical investigator remains an issue that we must address.

Second is an issue that no one really likes to talk about, and so I was pleased to see it listed: clinical investigators often are second-class citizens. The need for good to outstanding clinical investigators to work across disciplines, to be collaborative, to be the middle author on some publications, also is not discussed a lot. Our women's collaborative natures are, from my perspective, a benefit and a huge joy, but that particular aspect of our work is undervalued, and that needs to be addressed and corrected.

Now I want to move to Goldstein and Brown and a publication that I recommend to all of you and particularly love. In June 1997 Drs. Goldstein and Brown of cholesterol and Nobel laureate fame published in the *Journal of Clinical Investigation* a paper entitled "The Clinical Investigator: Bewitched, Bothered, and Bewildered but Still Beloved" that describes clinical investigators and clinical investigation.[2] It is a joy to read. According to Dr. Goldstein, these investigators are bewitched by the thrills of science (meaning laboratory-based investigation) and medicine (meaning the care they provide to patients), bothered by the need to choose one over the other, and bewildered by the need to choose.

They further posed a diagnosis that they called PAIDS. I've always thought of PAIDS as "pediatric acquired immunodeficiency syndrome." But Drs. Goldstein and Brown used PAIDS to refer to the "paralyzed academic investigators disease syndrome." They posited many of the same issues that have been discussed previously.

First, there is a knowledge gap between medical school and a research career, which in the context of this article is a career in clinical investigation. The knowledge gap is that a set of tools is required for laboratory-based investigation. Few of us have admitted until recently that an equally complex set of tools is demanded for the conduct of clinical investigation. Again, the National Institutes of Health are to be commended for their K-30 and K-12 awards, which are allowing some of our institutions to provide training in this area. Now we have to give our junior faculty release time to acquire these skills.

We all know that it is difficult to combine research with medical practice. Yet the conduct of clinical investigation implies direct interaction with patients, and it implies ongoing clinical practice. But then there are the economic disincentives. The debt load averages $80,000 for everybody who finishes medical school. But that's crept up, and by now it's probably over $100,000.

Drs. Goldstein and Brown say that four P's can be applied to patient-oriented research, or POR. The first is *Passion*—all of us are fairly passionate about our clinical investigation. The second is the *Patients* we care for and the patients for whom we define new therapies within our clinical investigation. The third is *Patience*. We've heard today that it takes a long time for a specific question posed within the context of clinical investigation to come to fruition. Goldstein and Brown say it took them eight years before they had anything meaningful they could translate to patients. And the fourth is *Poverty*. This does not mean personal poverty. It means grant dollar poverty, which has been somewhat remedied.

Moving Women into Leadership Positions

We know that women face many more challenges than men in obtaining career-advancing mentoring. We also know that men have difficulty effectively mentoring women. Men need some education, and, from my position, women do

---

[2] J. L. Goldstein and M. S. Brown, "The Clinical Investigator: Bewitched, Bothered, and Bewildered but Still Beloved," *Journal of Clinical Investigation* 99 (1997): 2803–2812.

too. There are just as many poor women senior mentors as there are poor men senior mentors. Senior women often are not leading the same lives and have not led the same lives as the junior women coming along. Senior women must then—and so must men—be very open and receptive to the changes in society and the increasing complexity of society if they're going to be good mentors.

Isolation reduces a woman's capacity for risk-taking, often translating into a reluctance to pursue professional goals or a protective response such as perfectionism. Why is risk-taking important, and why is it mentioned here, and why will I mention it later? I have a personal view that we make progress through taking risk. We make progress in terms of enhancing women's careers by taking risks at our institutions and nationally. We make progress in our research by going out on a ledge, asking questions that are not easy to answer. It's difficult to take risk if one is in isolation. If one has peers next door, down the hall, or in my case across the country to communicate with, it's easier to take risk. I find that although I'll be taking a risk, I can check with someone else that it's not necessarily a stupid risk. It's a good risk, a risk that will energize me and others, and a risk that will result in change and innovation. When women or men are isolated at any level—then risk-taking is diminished.

Without being conscious of their mental models of gender, both men and women still tend to devalue women's work and to allow women a narrower band of assertive behavior. The AAMC Increasing Women's Leadership Project Implementation Committee examined four years of school-provided data. Every institution receives a benchmark survey that's compiled annually by the AAMC. It includes interviews with department chairs across the country, both in clinical and in basic science departments, and it includes new research from industry and higher education on women's achievements and advancement.

The key findings are that women comprise 14 percent of tenured faculty, 12 percent of full professors, and 8 percent of department chairs in the United States and that very few schools, hospitals, or professional societies have a critical mass of women leaders. Moreover, the pool of women from which to recruit academic leaders remains shallow. I have to say that every time I present this, I'm told by my male colleagues to stop using the word *shallow*. Shallow can be interpreted in many ways. The pool does not necessarily include shallow women; it's just that the pool is not very deep.

How can we fix that? I believe we can fix it by working together and by generating a centralized pool that's easily accessible, so that when a great position is available, we can propose to the committee a group of women candidates who are serious candidates. But that list of candidates should not come just from our own list of preferences, but from a central pool of women who have said they are interested, they are prepared, they are academically viable.

According to AAMC data from 2001, 46 percent of new entrants to U.S. medical schools were women, and women were the majority of new entrants at 31 schools. At the level of residents or postdoctoral fellows, 38 percent of residents were women, with the highest proportion, 70 percent, as you might guess, in obstetrics and gynecology and 65 percent in pediatrics. A much lower proportion of women were in most surgical subspecialties such as urology, 12 percent; neurosurgery, 10 percent; orthopedic surgery, 8 percent; and thoracic surgery, 6 percent. So there's an uneven distribution of women entering the subspecialties and much work is left to be done. At the faculty level, 28 percent of all faculties were women, but only 12 percent of full professors were women.

The AAMC graph presented earlier by Janet Bickel (see Figure 3, p. 39) shows that the slope for medical school graduates has increased fairly consistently since about 1980. There are some little blips, but they're not substantial. On the other hand, the percentage of women faculty, which is now 28 percent, went up fairly rapidly during the late 1980s and the early 1990s, but it has leveled off and the slope is not nearly as steep as it was. To analyze the slope of that curve and the alteration in it, we need raw data that neither I nor the AAMC have.

And where are all the women faculty positioned? In 2001, 11 percent of all women faculty were full professors, 19 percent were associate professors, and by far the majority were at the beginning of their careers—50 percent of women faculty were assistant professors and over 17 percent were instructors (see Figure 4, p. 40).

So where should the academic societies focus their energy? If we're going to move women into leadership positions—meaning deans, chairs, division chiefs— we probably want to be working at the juncture between associate and full professors (Figure 4). But if we want to plan for the long haul, in a decade from now, we want to be working at the transition from assistant to associate professors. That's a charge I would make to the societies—to nurture, mentor, and work with the junior faculty in order to bring them along and be as certain as we can be that they'll be successful.

The data from my own school are not a lot different from anyone else's (Figure 5). The real reason for showing this is that we mark it against the AAMC benchmark data every year so that our chairs and deans are very aware of where we are doing well and where we are not doing so well. In 2001, 327 of our 605 medical students, or 53 percent, were women. We're very proud of this number. On the other hand, only 33 percent of our faculty, 486 out of 1,460, were women. This is higher than the AAMC data, but not good enough for an institution that's put a fair amount of resources into bringing women into the system and nurturing them, retaining them, and advancing them.

Where are these women at UCSF? Of all the full professors, 22 percent are women (Figure 6). Like on the AAMC national grid, the majority of women faculty at UCSF's School of Medicine are assistant professors or instructors, and, of course, the largest percentage are medical students. But let me emphasize that the research faculty are there. At UCSF only 125 of the 1,400 School of Medicine faculty are in a clinician educator series; about 700 are in the clinical and adjunct series (Figure 7). Few women are in the prestigious series; they sit in the less

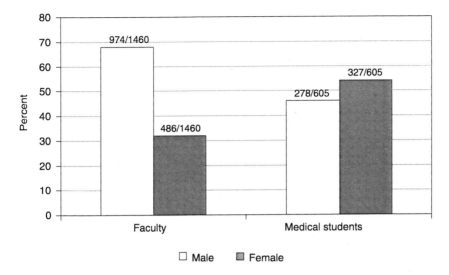

FIGURE 5 Gender distribution: full-time faculty versus medical students, University of California–San Francisco School of Medicine, 2001.
SOURCE: Unpublished data of the author.

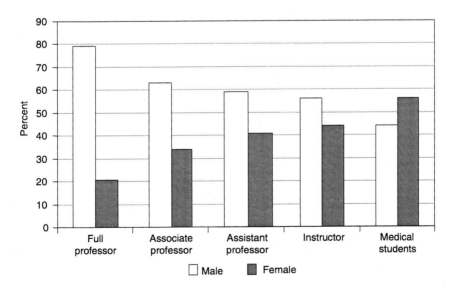

FIGURE 6 Gender distribution within faculty ranks and medical students, University of California–San Francisco School of Medicine, 2001.
SOURCE: Unpublished data of the author.

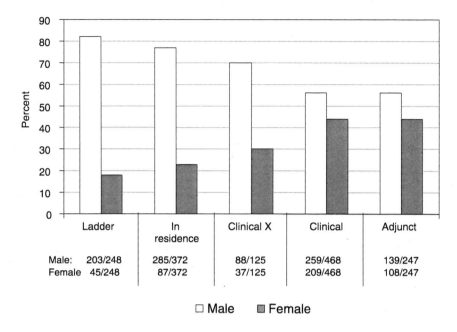

	Ladder	In residence	Clinical X	Clinical	Adjunct
Male:	203/248	285/372	88/125	259/468	139/247
Female	45/248	87/372	37/125	209/468	108/247

□ Male   ▨ Female

FIGURE 7 Gender distribution within series, University of California–San Francisco School of Medicine faculty, 2001.
SOURCE: Unpublished data of the author.

prestigious series. We at UCSF are not unique, but now we have women in the assistant professor and instructor ranks in the less prestigious series. Although I would love to be able to say five years from now that these are not less prestigious series, I don't believe we're going to move everyone in this direction. What we do need to do is value everyone for what they contribute.

Our initiatives to support women's progress have included mentoring programs. We have a mentorship faculty handbook, and we have a mentor-of-the-year award. What we have not done well is to develop family-friendly policies. Sue Shafer just co-chaired a climate survey at UCSF, which showed very clearly that both men and women were quite reasonably happy, over 60 percent happy, with their lives as academics at UCSF. But they were not happy about the absence of what are now termed family-friendly policies, including child-bearing and child-rearing policies, and the paucity of part-time positions "with a future" (one can work part time at UCSF but one doesn't necessarily have a future). Our faculty members also don't like that we do not have a day that actually begins and ends—our day is just a continuous cycle of 24 hours. And we don't have adequate elder care, which was sorted out in the climate survey and was noted as a weakness.

Although we do, I believe, have salary equity, it has not been institutionalized. Throughout this workshop I've heard both junior and senior faculty commenting on the perception that women are not paid equitably. To prove or disprove that and to make change, we have to institutionalize salary equity, meaning a database that can be checked twice a year with a variety of controls in place. We have to focus on retention, and we need to change our search committees for leadership and for all positions.

As for search committees, we need to bring women to the table. We have to get our leaders and search committees ready for women, and that means beyond tokens. We have to improve the search process, which in my mind means going beyond the research CV, increasing the emphasis on diversity, placing two to three women on each search committee. That implies all of us have to do more work, and when we're asked to serve on a search committee, we must say yes if we agree with this premise.

Finally, we need to avoid the reflecting pool method of hiring—that is, recruiting, hiring, and putting people on the short list who look just like the men on the search committee or perhaps even like me on the search committee. We need to look for people who have new ideas, who have new visions, and who are different from us. We have to request the reasons when the short list for any search committee does not include women or minority candidates.

Personal Perspectives

Clinical investigators are hybrids. I mentioned that they require skills both in the laboratory and in the clinic. They bring two worlds together and through their research give enhanced meaning to laboratory findings. What good is the human genome—all that work, all that money—if it's not correlated within the next five to ten years with the phenotype? Do my patients care if I can't explain to them the genetic basis of their disease or explain to them why a particular drug works differently for them than it does for the next person? So we need to move to importance, which is what has been done so beautifully in the laboratory over the last decade.

However, this clinical work implies patient demands. Night call and week-end call—it's another layer of complexity. I have no solution for this, by the way. Both academic medicine and society at large are much more complicated now than 20 years ago. I actually believe that raising my two children, who are now 29 and 33, was a lot easier than the job given to my young women colleagues.

Finding time to sustain and build both one's personal life and one's professional life is increasingly difficult. Speaking for myself and for my family, there are not enough hours in the day to care for aging parents (I cared for four of them over the last decade), to retain the solid friendships outside of academic medicine, to have space—here's that risk-taking thing again—for the necessary risk-taking that makes it all fun, to build skills (I like to think I'm still building) in clinical investigation, and to mentor others. That's a large platter, and there really needs to be room for all or at least most of it.

Then there's my daughter's wedding. She is working for Northern California Kaiser in obstetrics/gynecology, and so she has a lot of night call and weekend call. Because she's getting married next month, she and I talk every night about how we're going to divide up the jobs, because both of us have these very busy schedules. So it doesn't end; it just keeps going on and on. It's joyful and delightful, but there are problems for women that I believe are not there for men in balancing all of this.

The three-legged academic school of research, teaching, and community service does have a fourth leg: clinical responsibility for clinical investigators. One more thing, it's lonely at the top. As the MIT report on the status of women faculty said, and I thoroughly agree with it, marginalization does increase as women progress. At every institution there are too few senior women clinical scientists to form an effective cohort. So my mentors and my colleagues are people I work with nationally. I met them and work with them continually through the pediatric AIDS clinical trials group, through the pediatric primary immunodeficiency network, and through my societies.

Societies

What can societies do? Societies should be a forum for leadership. They should develop leaders

- through appointing women and men committee chairs, but they need to focus on women—they must
- through achievement awards—we should be nominating everyone who's eligible for research and achievement awards
- through session chairs—we need to be certain that women are session chairs and that they're junior women, because we want to nurture that large pool of women who are at the entry level
- through more data and best practices.

For career development, I like the notion of mentoring and mentoring awards. I like the notion of working with advancing women, of workshops for women within the context of society meetings. I especially like the notion that we should provide a mechanism for interaction between mid- and senior-level women to maintain ongoing networks. It's not just junior women who require the networking; it's all the way up to the top.

One society, not a clinical society, that I'm actually a member of, as are many people in this room, has done quite a good job. The American Society for Cell Biology [ASCB] has had women in the cell biology group since the mid- to late 1990s. The committee has a whole panoply of interactive and interesting pieces on the ASCB Web site [www.ascb.org], including AXXS '99. And they have a monthly newsletter that has become a standard at UCSF. Our active women in cell biology members forward it to us by e-mail, so that we'll all pay attention to it.

Today is an ideal time for societies to work toward the enhancement of career advancement for women in academic medicine and for women in clinical investigation, because the current environment, through the NIH, is providing unique support for young clinical investigators. It really positions professional societies to enhance women's careers in a way that I have not seen since the early 1970s. The dollars are there, and the women are in the door as assistant professors. It's a unique moment, and we should capitalize on it and perhaps just take a little risk and go ahead and do it.

Closing Remarks

Vivian Pinn, M.D.
Director, Office of Research on Women's Health
Associate Director, Research on Women's Health
National Institutes of Health

I want to begin by thanking Sally Shaywitz and all of you for your participation and helping to make this workshop successful. Many viable, innovative, and wonderful suggestions have come forward. Perhaps the most exciting results of this meeting, something that I enjoy every time I go to a meeting like this one, are the new interactions, new friendships, and new avenues for networking that have been established. Our speakers have all been marvelous.

As I listened to the recommendations, I liked hearing what we've all recognized—that societies can be and should be agents for change, and we want to facilitate that. We want them to be agents of change in the professions, in health careers, and in research careers for the entry and advancement of women—not only in the basic sciences, not only in medical sciences, but also in the clinical sciences. If we focus on our interdisciplinary collaborations, interdisciplinary research, and interdisciplinary career development, we will continue to ensure that our efforts are interdisciplinary in nature, spreading across not just the medical discipline but all of the other clinical disciplines, as well as basic science and perhaps traditional and nontraditional other areas through which women contribute both as basic scientists and as clinical scientists.

I liked some of the specific recommendations that were put forward. I heard three major things. The first was how to facilitate professional society meetings as a way to bring to the floor specific recommendations that would address the women who are members of those societies.

Second, as Dr. Pardes put forward, we need to bring together a panel on clinical research, including deans of medical schools and heads of medical centers, representatives of the NIH, and representatives of industry. We have heard people speak about the role of industry, but how do we add industry to the mix in terms of facilitating the advancement of women in biomedical research careers? Also, how do we get professional societies involved?

We will work with the Committee on Women in Science and Engineering and the National Academy of Sciences, because we feel this is a wonderful way to undertake a meeting that both has credibility and puts in correct focus the particular topic we want to address. So we will work very closely to see if we can fund just such a meeting in the coming year. It is hoped that we can take all the recommendations contained in the report of this meeting and then see where we can go in terms of bringing together the leaders in the research community. It's not just societies that are the agents of change, but also those who are in positions to be agents of change within the academic community, the academic health center community, the professional society community, and the NIH community.

Third is something that I thought was very important: being able to track progress, as Dr. Shaywitz and many others mentioned. We've heard accounts of individual institutions and individual societies. But how can we measure whether these activities and your participation are making any difference in your own professional societies? We've heard reference to things such as report cards; in fact, we heard that at AXXS '99. So should there be report cards for professional societies? Or are there better ways or other ways to track the impact of these kinds of meetings and discussions on efforts to increase activities to support women scientists among the different professional organizations?

I don't have the answer, and I didn't hear the answer from you. I did hear a call for a way to do that. So that's something else we need to pursue—how to better track the impact of these kinds of discussions. I have learned from my experience, and I'm sure you'd agree, that if there's a way to track the progress, there's more apt to be progress—that is, when someone knows that you're going to look closely for accountability about what has occurred.

Mentoring is something that again is central to almost everything that our office continues to focus on. I also have some ideas about many of the recommendations that I heard today, including looking at how to gain a forum at professional society meetings. I don't know that we could fund every society meeting that has a forum or a session on grant writing for women or on how to promote women's careers, but we ought to be able to do that for some number. I'm going to ask our careers committee to take a look at what we can do. We have funded a small number in the past, and perhaps we can come up with some way to at least be able to support some societies.

In closing, I want to thank you by making a commitment to take forward as best we can as many of the recommendations that you made that are appropriate for us to fund or that are appropriate for us to pursue. I promise you we will do that in gratitude for your effort.

Appendixes

Appendix A

Workshop Agenda

Achieving XXcellence:
The Role of Professional Societies in Advancing
Women's Careers in Science and Clinical Research
July 8-9, 2002

Sponsored by the Office of Research on Women's Health, National Institutes of Health, and the Burroughs Wellcome Fund

Monday, July 8 6:00–8:00 p.m.: Auditorium, National Academy of Sciences

6:20 p.m.   Welcome: Vivian Pinn, M.D., Associate Director for Research on Women's Health, Director, Office of Research on Women's Health, NIH

6:30 p.m.   Welcome: Sally Shaywitz, M.D., Chair, AXXS Steering Committee

6:45 p.m.   Carola Eisenberg, M.D., Harvard Medical School

7:00 p.m.   Refreshments

Tuesday, July 9 9:00 a.m.–4:30 p.m.: National Academy of Sciences

8:00 a.m.   Continental breakfast

9:00 a.m.   Opening remarks and introductions: Vivian Pinn, M.D.

9:15 a.m.   Chair, AXXS Steering Committee: Sally Shaywitz, M.D., Yale University School of Medicine

Welcome: Harvey Fineberg, M.D., Ph.D., President, Institute of Medicine

9:30 a.m.   Opening keynote: Karen Antman, M.D., Columbia Presbyterian Medical Center

10:00 a.m. From AXXS '99 to AXXS 2002: Page Morahan, Ph.D., National Center of Leadership in Academic Medicine

10:15 a.m. Break

10:30 a.m. A Pathways Model for Career Progression in Science: Pam Marino, Ph.D., National Institute of General Medical Sciences, NIH

10:45 a.m. Status Report and Recommendations on Advancing Women in Academic Medicine: Janet Bickel, M.A., Association of American Medical Colleges

11:00 a.m. Panel: Differences between Basic and Clinical Disciplines: W. Sue Shafer, Ph.D., Institute for Quantitative Biomedical Research; Herbert Pardes, M.D., New York Presbyterian Hospital; Jeanne Sinkford, D.D.S., Ph.D., American Dental Education Association; Moderator: Sally Shaywitz, M.D.

11:45 a.m. Ruth Kirschstein, M.D., Deputy Director, NIH

12:00 p.m. Working lunch, breakout rooms

12:30 p.m. Breakout sessions: During this session, participants will discuss implementation of action plans within their societies for developing support and launching programs to advance women.

2:00 p.m.   Break

2:15 p.m.   Breakout sessions continued

3:00 p.m.   Prepare reports from breakout sessions

3:30 p.m.   Plenary: reports from breakout sessions

3:45 p.m.   Closing plenary: Diane Wara, M.D., University of California–San Francisco

4:15 p.m.   Closing remarks

4:30 p.m.   Adjourn

Appendix B

Workshop Participants and Speakers

Participants

Nahrain Alzubaidi, M.D.
Endocrine Society

Joan Amatniek
Director, CNS Medical Affairs, Neurology
Janssen Pharmaceuticals

Debra Babcock
Medical Officer
National Institute of Mental Health (NIMH/NIH)

Rebecca Bahn
Consultant in Endocrinology
Endocrine Society

Angela Bates
NIH Office of Research on Women's Health

Lisa Begg, Dr.P.H.
Director of Research Programs
NIH Office of Research on Women's Health

Susan Benloucif, Ph.D.
Research Associate Professor
Dept. of Neurology
Northwestern University Medical School

Mary Berg
Professor
Society for Women's Health Research

Diane Bernal
Director, Intramural Management, NEI/DIR
NIH Office of Research on Women's Health

Maria P. Bettinotti, Ph.D., dipl. ABHI
Research Scientist, Molecular Immunology Laboratory
NIH Department of Transfusion Medicine
Clinical Center

Arlene Bierman, M.D., M.S.
Senior Research Physician
American Geriatrics Society

Keri Biscoe, M.D.
National Eye Institute, Lab. of Sensorimotor Research

Joann Boughman, Ph.D.
Executive Vice President
American Society of Human Genetics

Mary Bouxsein, Ph.D.
American Society for Bone and Mineral Research

Beth Bowers
Social Science Analyst
Division of Services & Intervention Research
NIMH, NIH

Leslie Cameron
Women's Programs Officer
American Psychological Association

Deborah Carper, Ph.D.
Chair, NIH Women Scientist Advisory Council
National Eye Institute, NIH

Leticia Castillo, M.D.
Society of Critical Care Medicine

Manisha Chandalia, M.D.
UT Southwestern Medical Center

Zhong Chen, M.D., Ph.D.
Staff Scientist
American Academy of Otolaryngology-Head and Neck Surgery

Valarie Clark
Associate Director, Women's Progs/Faculty Affairs
Association of American Medical Colleges

Thomas Crist
Policy Associate
Association of Professors of Medicine

Joan C. Davis, M.D., M.P.H.
NICHD/RSB

Mary Lou de Leon Siantz
President
National Association of Hispanic Nurses

Donna J. Dean
Acting Director
NIBIB/NIH

Catherine Didion
Executive Director
Association for Women in Science

Ilise L. Feitshans, J.D., Sc.M.
New Jersey Developmental Disabilities Council

Rose Fife, M.D.
American Medical Women

Loretta Finnegan
Medical Advisor
Society for Pediatric Research

Maryrose Franko
Senior Program Officer
Howard Hughes Medical Institute

Gail Gamble, M.D.
American Academy of Physical Medicine and Rehabilitation

Lynn Gerber
Chief, Rehabilitation Medicine Dept
American Academy of Physical Medicine and Rehabilitation

Marise Gottlieb, M.D.
American Federation for Medical Research

Kelly Gull
Manager, Education Programs
American Society for Biochemistry and Molecular Biology

Alan Guttmacher, M.D.
Deputy Director
National Human Genome Research Institute

Carol Haggans
Program Analyst
NIH Office of Dietary Supplements

Susan Hahn
CPA, EMT
KPMG

Eleanor Hanna
Associate Director for Special Projects and Centers
NIH Office of Research on Women's Health

Florence Haseltine
Director
NIH Center for Population Research

John Hawley
Executive Director
American Society for Clinical Investigation

Roberta L. Hines, M.D.
Professor and Chairman
Department of Anesthesiology
Yale University School of Medicine

Andrea Hoberman
Post-Baccalaureate Intramural Research Training Award (IRTA) Fellow
NIMH, Mood and Anxiety Disorders Program

Alice Hogan
Program Director
National Science Foundation

Katherine Hollinger
Senior Health Promotions Officer
Food and Drug Administration Office of Women's Health

Kimberly Howell
Office of the Director
National Institute on Aging

Sharon Hrynkow, Ph.D.
Deputy Director
NIH Fogarty International Center

Michelle A. Josephson, M.D.
American Society of Transplantation

Miriam F. Kelty, Ph.D.
Associate Director
National Institute on Aging

Mahin Khatami, Ph.D.
President
Graduate Women In Science

Sooja Kim
Chief, Endo/Metb/Nutr/Reproductive Sciences IRG
NIH Office of Research on Women

Sheri M. Krams, Ph.D.
American Society of Transplantation

Mary Lawrence
Program Analyst
NIH Office of Research on Women

Evelyn Lewis&Clark
Associate Chair for Research, Dept of Fam Med
Society of Teachers of Family Medicine

Mary M. Lewis, R.N., B.S.N.
Research Nurse Specialist
National Cancer Institute

Martha Liggett, Esq.
Executive Director
American Society of Hematology

Estelle Lin
NIH Office of Research on Women's Health

Irene Litvan, M.D.
Chief, Cognitive Neuropharmacology Unit
American Neurological Association

Joan K. Lunney, Ph.D.
Research Leader, USDA
American Association of Immunologists

Linda Mah
Clinical Fellow
American Psychiatric Association

Vicki L. Malick
Health Education Analyst
Office of Intramural Research
Intramural Program on Research on Women's Health (IPRWH)

Mary A. Marovich, M.D., DTMH
HIV Vaccine Development

Sherry Marts
Scientific Director
Society for Women's Health Research

Erin McClure, Ph.D.
Post-Doctoral Fellow
NIH NIMH/MAP

Elizabeth McGregor
Associate Director
Institute on Gender and Health (IGH)

Deborah McPherson, M.D.
Asst Dir, Medical Educ
American Academy of Family Physicians

Michael Milano
Training and Organizational Development Consultant
ACTeam (AXXS Coordinating Team)

Mojdeh Moghaddam, Ph.D.
Neuroscientist Society for Neuroscience

Gwen Myers, Ph.D.
Staff Liaison for Research Committee
Association of Academic Physiatrists

Barbara M. Myklebust, Ph.D.
Research Scientist
The George Washington University

Carol Nicholson, M.D., M.S., FAAP
Program Director
NIH/NICHD/NCMRR/PCCR
National Institutes of Health/NICHD

Serene Olin, Ph.D.
NIMH

Jessica S. Parker, M.S.
NIDCD

Delores L. Parron, Ph.D.
Scientific Advisor for Capacity Development
Office of the Director, National Institutes of Health

Estella Parrott, M.D., M.P.H.
Program Director, NICHD
NIH Office of Research on Women's Health

Ellyn Pollack, M.A., APR
Communications Director
NIH Office of Research on Women's Health

Fareen Pourhamidi, M.S., M.P.H.
Research Resource Coordinator
American Academy of Otolaryngology - Head and Neck Surgery Foundation

Peggy A. Pritchard, M.S.
American Society for Microbiology

Margaret V. Ragni, M.D.
Director, Hemophilia Center of Western PA
American Society of Hematology

Jenny Read, Ph.D.
National Eye Institute
Lab. of Sensorimotor Research

Rosalyn Richman
Co-Director, ELAM Program
Executive Leadership in Academic Medicine (ELAM) Program

Joyce Rudick
Director, Programs and Management
NIH Office of Research on Women's Health

Jennifer Saxman-Tesfaye
Admin Asst
NIH, NHGRI

Joannie Shen, M.D., Ph.D.
Society for Neuroscience

Janine Smith
Deputy Clinical Director
National Eye Institute

Mark Sobel
Executive Officer
American Society for Investigative Pathology, Inc.

Roberta Sonnino, M.D.
Associate Dean, Women in Medicine and Special Programs;
Professor of Pediatric Surgery
Association of Women Surgeons

Anita Miller Sostek
Chief
NIH Behavioral and Biobehavioral Processes IRG, CSR

Esther Sternberg, M.D.
NIH Office of Research on Women's Health

Joanita Stokes
Psychology Student, University of Maryland
American Psychological Association

Nancy Sung
Program Officer
Burroughs Wellcome Fund

Eva Szigethy, Ph.D.
American Psychiatric Association

Sandra Talley
NIH National Heart, Lung, and Blood Institute

Har Tan
Summer Intern
National Institutes of Mental Health, Mood and Anxiety Disorder Program

Kimberly Templeton
Treasurer
Ruth Jackson Orthopaedic Society

Nancy Thompson, Ph.D.
Professor of Medicine and Pathology
American Society for Investigative Pathology, Inc.

Shirley Tom Chan
Pharm.D.
Johnson & Johnson

Joyce Townser, R.N., B.S.N.
Region VII Women's Health Coordinator
US Public Health Service

Esa Washington, M.D., MPH
Society of Teachers of Family Medicine

Rudolph Williams
Executive Director
National Medical Association

Joyce Woodford
Minority and Women's Health Program Officer
National Institute of Allergy and Infectious Diseases

Annette B. Wysockim, Ph.D.
Wound Healing Society

Parvin M. Yasaei, Ph.D.
Food and Drug Administration
Center for Food Safety and Applied Nutrition

Speakers

Karen Antman, M.D.
Wu Professor of Medicine and Professor of Pharmacology
Columbia Presbyterian Medical Center

Carola Eisenberg, M.D.
Lecturer on Social Medicine
Harvard Medical School

Harvey Fineberg, M.D., Ph.D.
President
Institute of Medicine

Ruth Kirschstein, M.D.
Deputy Director
National Institutes of Health

Pamela Marino, Ph.D.
Program Director
National Institute of General Medical Sciences, NIH

Page S. Morahan, Ph.D.
Founding Director, National Center of Leadership in Academic Medicine
American Society for Microbiology

Vivian Pinn, M.D.
Director
NIH Office of Research on Women's Health

Diane Wara, M.D.
Division Chief of Pediatric Immunology/ Rheumatology
UCSF School of Medicine

Steering Committee

Nancy Andrews, M.D., Ph.D.
Leland Fikes Professor of Pediatrics
Harvard Medical School

Michael Lockshin, M.D.
Professor of Medicine
Weill College of Medicine of Cornell University

Janet Bickel, M.A.
Associate Vice President for Medical School Affairs
Association of American Medical Colleges

Deborah Powell, M.D.
Executive Dean and Vice Chancellor
     for Clinical Affairs
University of Kansas School of Medicine

Herbert Pardes, M.D.
President and CEO
New York Presbyterian Hospital

W. Sue Shafer, Ph.D.
Deputy Director
Institute for Quantitative Biomedical Research

Jeanne Sinkford, D.D.S., Ph.D.
Associate Executive Director
The American Dental Education Association

Sally Shaywitz, M.D.
Professor of Pediatrics
Yale University School of Medicine

Clinical Research Roundtable Liaison

Veronica Catanese, M.D.
Senior Associate Dean for Medical Education
New York University Medical Center

NRC Staff

Jong-on Hahm, Ph.D.
Director
Committee on Women in Science and Engineering

Amaliya Jurta
Senior Program Assistant
Committee on Women in Science and Engineering